SpringerBriefs in Electrical and Computer Engineering

Computational Electromagnetics

Series editor

Rakesh Mohan Jha, Bangalore, India

More information about this series at http://www.springer.com/series/13885

Balamati Choudhury · Pavani Vijay Reddy
Rakesh Mohan Jha

Permittivity and Permeability Tensors for Cloaking Applications

 Springer

Balamati Choudhury
Centre for Electromagnetics
CSIR-National Aerospace Laboratories
Bangalore, Karnataka
India

Rakesh Mohan Jha
Centre for Electromagnetics
CSIR-National Aerospace Laboratories
Bangalore, Karnataka
India

Pavani Vijay Reddy
Centre for Electromagnetics
CSIR-National Aerospace Laboratories
Bangalore, Karnataka
India

ISSN 2191-8112   ISSN 2191-8120   (electronic)
SpringerBriefs in Electrical and Computer Engineering
ISSN 2365-6239   ISSN 2365-6247   (electronic)
SpringerBriefs in Computational Electromagnetics
ISBN 978-981-287-804-5   ISBN 978-981-287-805-2   (eBook)
DOI 10.1007/978-981-287-805-2

Library of Congress Control Number: 2015947807

Springer Singapore Heidelberg New York Dordrecht London

Printed on acid-free paper

Springer Science+Business Media Singapore Pte Ltd. is part of Springer Science+Business Media (www.springer.com)

To Professor R. Narasimha

In Memory of Dr. Rakesh Mohan Jha
Great scientist, mentor, and excellent
human being

Dr. Rakesh Mohan Jha was a brilliant contributor to science, a wonderful human being, and a great mentor and friend to all of us associated with this book. With a heavy heart we mourn his sudden and untimely demise and dedicate this book to his memory.

Preface

The advancement in metamaterial science and technology has led to some of the interesting applications over the entire frequency spectrum. One among these exciting applications is the invisibility cloak. The concept of invisibility has attracted researchers in the aerospace discipline to implement it for low observability. The Maxwell's equations in conjunction with transformation optics in anisotropic media is the fundamental principle of operation of invisibility cloak. A major challenging issue in the design of invisibility cloak lies in the determination of permittivity and permeability tensors for each cloaking shell.

This brief systematically describes analytical expressions for the permittivity and permeability tensors for various quadric surfaces that find application in aerospace platforms. The spatial metric derivations for permittivity tensors of circular, elliptic, hyperbolic and parabolic cylinders, sphere, right circular cone, prolate spheroid, oblate spheroid, general paraboloid of revolution, and ogive have been included in this brief. Various cloaking shells of arbitrary shapes can be designed with the help of this mathematical formulation of diverse quadratics and their hybrids.

Balamati Choudhury
Pavani Vijay Reddy
Rakesh Mohan Jha

Acknowledgments

We would like to thank Mr. Shyam Chetty, Director, CSIR-National Aerospace Laboratories, Bangalore for his permission and support to write this SpringerBrief.

We would also like to acknowledge valuable suggestions from our colleagues at the Centre for Electromagnetics, Dr. R.U. Nair, Dr. Hema Singh, Dr. Shiv Narayan, Dr. Balamati Choudhury, and Mr. K.S. Venu during the course of writing this book.

But for the concerted support and encouragement from Springer, especially the efforts of Suvira Srivastav, Associate Director, and Swati Meherishi, Senior Editor, Applied Sciences & Engineering, it would not have been possible to bring out this book within such a short span of time. We very much appreciate the continued support by Ms. Kamiya Khatter and Ms. Aparajita Singh of Springer towards bringing out this brief.

Contents

About the Authors

Dr. Balamati Choudhury is currently working as a scientist at Centre for Electromagnetics of CSIR-National Aerospace Laboratories, Bangalore, India since April 2008. She obtained her M.Tech. (ECE) degree in 2006 and Ph.D. (Engg.) degree in Microwave Engineering from Biju Patnaik University of Technology (BPUT), Rourkela, Orissa, India in 2013. During the period of 2006–2008, she was a Senior Lecturer in Department of Electronics and Communication at NIST, Orissa, India. Her active areas of research interests are in the domain of soft computing techniques in electromagnetics, computational electromagnetics for aerospace applications and metamaterial design applications. She was also the recipient of the CSIR-NAL Young Scientist Award for the year 2013–2014 for her contribution in the area of Computational Electromagnetics for Aerospace Applications. She has authored and co-authored over 100 scientific research papers and technical reports including a book and three book chapters. Dr. Balamati is also an Assistant Professor of AcSIR, New Delhi.

Ms. Pavani Vijay Reddy is currently working as Project Assistant at Center for Electromagnetics (CEM), CSIR-NAL, Bangalore. She obtained her B.Tech. degree from Jawaharlal Nehru Technological University, Kakinada, and she is working on conformal antennas, metamaterials, invisibility cloaking, etc.

Dr. Rakesh Mohan Jha was Chief Scientist & Head, Centre for Electromagnetics, CSIR-National Aerospace Laboratories, Bangalore. Dr. Jha obtained a dual degree in BE (Hons.) EEE and M.Sc. (Hons.) Physics from BITS, Pilani (Raj.) India, in 1982. He obtained his Ph.D. (Engg.) degree from Department of Aerospace Engineering of Indian Institute of Science, Bangalore in 1989, in the area of computational electromagnetics for aerospace applications. Dr. Jha was a SERC (UK) Visiting Post-Doctoral Research Fellow at University of Oxford, Department of Engineering Science in 1991. He worked as an Alexander von Humboldt Fellow at the Institute for High-Frequency Techniques and Electronics of the University of Karlsruhe, Germany (1992–1993, 1997). He was awarded the Sir C.V. Raman Award for Aerospace Engineering for the Year 1999. Dr. Jha was elected Fellow of INAE in 2010, for his contributions to the EM Applications to Aerospace

Engineering. He was also the Fellow of IETE and Distinguished Fellow of ICCES. Dr. Jha has authored or co-authored several books, and more than five hundred scientific research papers and technical reports. He passed away during the production of this book of a cardiac arrest.

Abbreviations

EM	Electromagnetic
GPOR	General Paraboloid of Revolution
QYACYLs	Quadric cylinders
QUASORs	Quadric surfaces of revolution
THz	Terahertz

List of Figures

Permittivity and Permeability Tensors for Cloaking Applications

Abstract An optimal version of electromagnetic (EM) stealth is the design of invisibility cloak of arbitrary shapes in which the EM waves can be controlled within the cloaking shell by introducing a prescribed spatial variation in the constitutive parameters. The promising challenge in design of invisibility cloak lies in the determination of permittivity and permeability tensors for all the layers. This book provides the detailed derivation of analytical expressions of the permittivity and permeability tensors for various quadric surfaces within the 11 Eisenhart co-ordinate systems. These include the cylinders and the surfaces of revolutions. The analytical modelling and spatial metric for each of these surfaces are provided along with their tensors. This mathematical formulation will help the EM designers to analyze and design various quadratics and their hybrids, which can eventually lead to the designing of cloaking shells of arbitrary shapes.

Keywords Invisibility cloak · Stealth · Permittivity tensors · Permeability tensors · Analytical modelling · Surfaces of revolutions

1 Introduction

The exciting features of metamaterials attracted the attention of researchers for various novel applications in different frequency regions those includes microwave, terahertz (THz) region, infrared, optics, acoustics, etc. The novel application is the invisibility cloak. The design of a invisibility cloak is an inverse problem where electromagnetic (EM) waves are controlled through a prescribed spatial variation in the material parameters (Schurig et al. 2006).

The promising challenge in design of invisibility cloak lies in the calculation of permittivity and permeability tensors at each and every layer as the electromagnetic wave has to bend inside the cloak in a desired manner. This can be achieved by calculating the constitutive parameters of bi-anisotropic materials through space time metric. This can lead towards the inverse formulation of flat space time to curved space time.

© The Author(s) 2016
B. Choudhury et al., *Permittivity and Permeability Tensors for Cloaking Applications*, SpringerBriefs in Computational Electromagnetics, DOI 10.1007/978-981-287-805-2_1

The invisibility cloak structure is dependent on the type of object to be cloaked (Choudhury et al. 2013), hence it is very much important to calculate the permeability and permittivity tensors in all the directions (uu, vv, and zz direction) w.r.t. the various 3-D quadric surfaces. Further these parameters can be optimized using the highly efficient soft computing techniques for design and analysis of the metamaterial invisibility cloak (Choudhury et al. 2012; Ren et al. 2011; Lee et al. 2012). Before describing the basic concepts of permittivity and permeability tensors along with their derivations to complex quadric surface of revolutions, a brief review of the work done in this area are presented here.

The first practical 2D metamaterial cloak was realized by (Schurig et al. 2006). This group, have used the conventional approach, where the permittivity and permeability tensors were derived by co-ordinate transformation while keeping the Maxwell's equations invariant in one co-ordinate system. The invisibility was imperfect due to approximations and material absorption. Further the permittivity and permeability tensors were derived for a cylindrical surface only. Leonhardt (2006) introduced a conformal mapping technique to reduce the imperfections in perfect invisibility within the accuracy of geometrical optics for objects that are larger than the wavelength of operation. This technique also uses the same conventional approach towards deriving permittivity and permeability tensors and proved that it is difficult to achieve perfect invisibility (because waves are not only refracted at the boundary but also reflected), but the reflection can be reduced by anti-reflection coating. Soon after few other groups used the same method and designed cloaks of various shapes (Cai et al. 2007; Xu et al. 2010).

Further (Zhang et al. 2008) used the constitutive parameters of bi-anisotropic materials through space time metric to calculate the corresponding permittivity and permeability tensors. This can lead towards the inverse formulation of flat space time to curved space time. This method was further used by Ahn et al. (2006) and they limited the derivations up to three structures like elliptical cylinder, confocal paraboloid, and prolate spheroid. The work presented here derives the spatial metric, permittivity and permeability tensors for all the structures that can be generated using the eleven Eisenhart coordinated systems. These generated structures can further be mapped to almost all aerospace structures which enhances the possibility of using cloaking effects towards stealth application. Further the permittivity tensors for an ogive (an example of extended orthogonal coordinated system) shaped cloaking shell is also derived and reported in this document.

2 Basic Concept of Permeability and Permittivity Tensors

The fundamental quantities like the dielectric constant and the permeability are the *a apriori* requirement for propagation analysis of electromagnetic waves in matter. For anisotropic random media these fundamental parameters acts like tensors. Hence in this section the basic concepts of permeability and permittivity tensors using effective medium approach (Leonhardt et al. 2006; Leonhardt et al. 2009) in general relativity is provided. The formulation has been extended for obtaining the relative parameters for various quadric shaped invisibility devices.

Many researchers are trying to make a device perfectly invisible and this can be achieved by guiding the electromagnetic wave around the object. This is based on the fact that Maxwell equations have the same form in all coordinate systems. The values of the electrical and magnetic field as well as the permittivity and permeability tensors are only changed by coordinate transformation.

There are two key problems in research of invisibility cloaking. The first one is calculation of the parametric equation of permittivity and permeability tensors for materials filled in the cloaking shell and the second one is the practical H/w realization of the metamaterial cloak (Ma et al. 2008).

Since the main focus is in the application of invisibility devices in the three dimensional space, spatial transformation media which perform spatial coordinate transformation of electromagnetic field has been considered.

Step 1-*Define the parametric equation of the quadric surface*: The cartesian coordinates x, y, and z are interrelated to the x^1, x^2 and x^3 by

$$x = x(x^1, x^2, x^3); \quad y = y(x^1, x^2, x^3); \quad z = z(x^1, x^2, x^3) \tag{1}$$

The object to be cloaked and the cloaking shell both are represented using curvilinear coordinate system (x^1, x^2, x^3). The dimensions of the cloaking object and the cloaking shell (Fig. 1) are given in Eqs. (2) and (3) respectively.

$$0 < x^1 < U_1; \quad 0 < x^2 < V_1; \quad 0 < x^3 < W_1; \tag{2}$$

$$U_1 < x^1 < U_2; \quad V_1 < x^2 < V_2; \quad W_1 < x^3 < W_2 \tag{3}$$

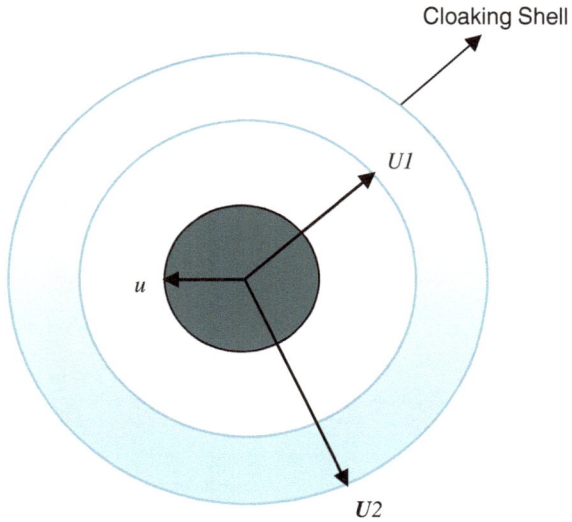

Fig. 1 Dimension of the object to be cloaked and the cloaking shell

Hence the arc length can be given as

$$ds^2 = dx^2 + dy^2 + dz^2 = \sum_{ij} \gamma_{ij} dx^i dx^j \tag{4}$$

For defining the physical medium for an empty curved space-time the primed coordinate system was used as given below:

$$x^1 = U_1 + \frac{U2 - U1}{U2} x'^1; \quad x^2 = V_1 + \frac{V2 - V1}{V2} x'^2; \quad x^3 = W_1 + \frac{W2 - W1}{W2} x'^3 \tag{5}$$

Step 2-*Derive the position vector for the quadric surface*: As mentioned above the transformation equations from cartesian coordinates (x, y, z) to curvilinear coordinates (x^1, x^2, x^3) as given in Eq. (1). The position vector for the curvilinear coordinated system is given as:

$$\vec{p} = \vec{p}(x^1, x^2, x^3) = x(x^1, x^2, x^3)\hat{e1} + y(x^1, x^2, x^3)\hat{e2} + z(x^1, x^2, x^3)\hat{e3} \tag{6}$$

and consequently

$$d\vec{p} = \frac{\partial \vec{p}}{\partial x^1} dx^1 + \frac{\partial \vec{p}}{\partial x^2} dx^2 + \frac{\partial \vec{p}}{\partial x^3} dx^3 \tag{7}$$

$$\vec{E1} = \frac{\partial \vec{p}}{\partial x^1} = \frac{\partial \vec{x}}{\partial x^1} \hat{e1} + \frac{\partial \vec{y}}{\partial x^1} \hat{e2} + \frac{\partial \vec{z}}{\partial x^1} \hat{e3} \tag{8}$$

$$\vec{E2} = \frac{\partial \vec{p}}{\partial x^2} = \frac{\partial \vec{x}}{\partial x^2} \hat{e1} + \frac{\partial \vec{y}}{\partial x^2} \hat{e2} + \frac{\partial \vec{z}}{\partial x^2} \hat{e3} \tag{9}$$

$$\vec{E3} = \frac{\partial \vec{p}}{\partial x^3} = \frac{\partial \vec{x}}{\partial x^3} \hat{e1} + \frac{\partial \vec{y}}{\partial x^3} \hat{e2} + \frac{\partial \vec{z}}{\partial x^3} \hat{e3} \tag{10}$$

where, $E1$, $E2$, $E3$ are tangent vectors to the quadric surfaces defined in curvilinear coordinate system.

Step 3-*Derive the spatial metric using position vector*: The components of the spatial metric can be defined as:

$$\gamma 11 = \frac{\partial p}{\partial x^1} \times \frac{\partial p}{\partial x^1}; \quad \gamma 12 = \frac{\partial p}{\partial x^1} \times \frac{\partial p}{\partial x^2}; \quad \gamma 13 = \frac{\partial p}{\partial x^1} \times \frac{\partial p}{\partial x^3} \tag{11}$$

$$\gamma 21 = \frac{\partial p}{\partial x^2} \times \frac{\partial p}{\partial x^1}; \quad \gamma 22 = \frac{\partial p}{\partial x^2} \times \frac{\partial p}{\partial x^2}; \quad \gamma 23 = \frac{\partial p}{\partial x^2} \times \frac{\partial p}{\partial x^3} \tag{12}$$

$$\gamma 31 = \frac{\partial p}{\partial x^3} \times \frac{\partial p}{\partial x^1}; \quad \gamma 32 = \frac{\partial p}{\partial x^3} \times \frac{\partial p}{\partial x^2}; \quad \gamma 33 = \frac{\partial p}{\partial x^3} \times \frac{\partial p}{\partial x^3} \tag{13}$$

where,

$$\gamma_{ij} = E_i \cdot E_j = \frac{\partial p}{\partial x^i} \times \frac{\partial p}{\partial x^j} \tag{14}$$

In this special metric, the components of $\gamma_{ij} = 0$ when $i \neq j$ (orthogonal systems). Hence the spatial metric can now be defined as:

$$\gamma_{ij} = \begin{pmatrix} h_1{}^2 & 0 & 0 \\ 0 & h_2{}^2 & 0 \\ 0 & 0 & h_3{}^2 \end{pmatrix} \tag{15}$$

where,

$$h1 = \overline{)|E1|}; \quad h2 = \overline{)|E2|}; \quad h3 = \overline{)|E3|}; \tag{16}$$

The effective geometry g_{ij} (Ahn et al. 2006) corresponding to the bi-anisotropic medium can be defined aswhere u is having range of

$$g_{ij} = \frac{\partial x^i}{\partial x'^k} \frac{\partial x^j}{\partial x'^l} \gamma'^{kl}; \tag{17}$$

The volume element of mathematical space $\sqrt{g}$ is given as

$$\sqrt{-g} = \sqrt{-g'} * \frac{\partial(x'^1, x'^2, x'^3)}{\partial(x^1, x^2, x^3)} = \sqrt{\gamma'} * \frac{\partial(x'^1, x'^2, x'^3)}{\partial(x^1, x^2, x^3)} \tag{18}$$

here we have

$$\gamma = \det(\gamma_{ij}); \tag{19}$$

Equations (13)–(17) give the transformation formula for the calculation of permittivity and permeability tensors. The constitutive parameters (considering the spatially covariant forms of divergences in the Maxwell's equations), are then given by (Leonhardt and Philbin 2006; Leonhardt and Tyc 2009; Plebanski 1960):

$$\varepsilon^{ij} = \mu^{ij} = \mp \frac{\sqrt{g}}{g_{oo}\sqrt{\gamma}}(g_{ij}) \tag{20}$$

The negative sign in Eq. (19) indicates the material as a negative refractive index material. As the EM wave guided around the cloaking shell, the object inside the boundary is hidden. This document focuses on deriving permittivity tensors, as the invisibility cloaking shell is made up of anisotropic media which can guide the EM wave around the object to be cloaked.

3 Permeability and Permittivity Tensor for Quadric Cylinders

Analysis and design of the cloak depends on the type of the object to be cloaked and its dimension. Hence, to find out the analytical expressions for cloaking parameters, the knowledge of analytical expressions are the a priori requirement. In this context the analytical parametric equations of the quadric cylinders along with the derivations for permittivity tensors are described in this section. There are different types of quadric cylinders (QYACYLs) such as right circular cylinder, right elliptic cylinder, right hyperbolic cylinder, and right parabolic cylinder etc.

3.1 Permittivity and Permeability Tensor for Circular Cylindrical Cloaking Shell

The simplest developable surface that is generally used for various electromagnetic applications is a right circular cylinder. Hence, the first structure considered for calculation of permittivity tensor is a right circular cylinder. The parametric equation (Jha et al. 1995) of a right circular cylinder is given as:

$$x = u \cos v; \quad y = u \sin v; \quad z = z \tag{21}$$

where, u is the radius of the cylinder, v is the angle which varies from $0°$ to $360°$ and z is the finite height of the cylinder. Figure 2 shows a typical right circular cylinder generated using Matlab which is the boundary of a cloaking shell.

The position vector for the circular cylindrical structure was derived using the above mentioned parametric equation (Appendix A) and the spatial metric is calculated as given below

$$\gamma_{ij} = \begin{pmatrix} 1 & 0 & 0 \\ 0 & u^2 & 0 \\ 0 & 0 & 1 \end{pmatrix}; \quad \gamma = \det(\gamma_{ij}) = u^2 \tag{22}$$

Assuming, the region where the object to be hidden $0 < u < U1$, and the region of invisibility cloaking shell is $U1 < u < U2$, then the physical medium w.r.t. the primed empty curved space-time is given by

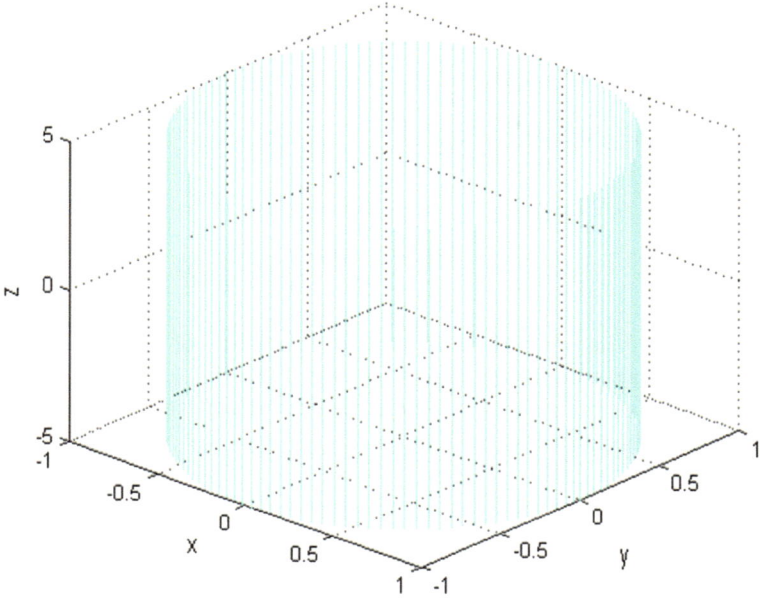

Fig. 2 Boundary of a right circular cylindrical cloaking shell

$$u = U1 + u' * \frac{(U2 - U1)}{U2}; \quad v = v'; \quad z = z'; \tag{23}$$

$$\frac{\partial u}{\partial u'} = \frac{(U2 - U1)}{U2} \tag{24}$$

The effective geometry corresponding to the bi-anisotropic medium can be defined as:

$$g_{ij} = \frac{\partial x^i}{\partial x'^k} \frac{\partial x^j}{\partial x'^l} \gamma'^{kl} = \begin{pmatrix} \left(\frac{U2-U1}{U2}\right)^2 & 0 & 0 \\ 0 & 1/u'^2 & 0 \\ 0 & 0 & 1 \end{pmatrix}; \tag{25}$$

$$\sqrt{-g} = \sqrt{-g'} * \frac{\partial(x'^1, x'^2, x'^3)}{\partial(x^1, x^2, x^3)} = u'\left(\frac{U2}{U2 - U1}\right); \tag{26}$$

$$\sqrt{\gamma} = u \tag{27}$$

The permittivity tensors are given by

$$\varepsilon^{ij} = \pm \frac{\sqrt{-g}}{\sqrt{\gamma}} * g^{ij} \tag{28}$$

Hence the permittivity tensor for a right circular cylindrical cloaking shell can be derived using Eqs. (25) through (27) as:

$$
\varepsilon^{ij} = \frac{\left(\frac{U2}{U2-U1}\right)u'}{u} * \begin{pmatrix} \left(\frac{U2-U1}{U2}\right)^2 & 0 & 0 \\ 0 & 1/u'^2 & 0 \\ 0 & 0 & 1 \end{pmatrix}
$$

$$
= \begin{pmatrix} \left(\frac{U2-U1}{U2}\right) & 0 & 0 \\ 0 & \left(\frac{U2}{U2-U1}\right) * \frac{1}{u'} & 0 \\ 0 & 0 & \left(\frac{U2}{U2-U1}\right) * u' \end{pmatrix} * \frac{1}{u} \tag{29}
$$

The permittivity tensor derived is used for simulation of typical right circular cylindrical cloaking shell having dimensions $u = 1$, $U1 = 1.1$, $U2 = 1.2$. To hide a cylindrical object of dimension $0 < u < U1$, the permittivity tensors in the uu, vv and zz directions are given in Fig. 3a through c.

3.2 Permittivity Tensor for Elliptic Cylindrical Cloaking Shell

Most of the structures in various platforms those including aerospace are either elliptic cylindrical or their hybrids in nature. Hence, the study of permittivity tensors for elliptic cylindrical structures is given here. The parametric equation of a right elliptic cylinder is given as:

$$
x = \cosh u * \cos v; \quad y = \sinh u * \sin v; \quad z = z \tag{30}
$$

where u is $(0 \leq \infty)$, v is the angle which varies from $0°$ to $360°$ and z is the finite height of the cylinder. Figure 4 shows the boundary of a right elliptic cylindrical cloaking shell inside which an object can be hidden.

The position vector of the above mentioned parametric equation was derived (Appendix B) and the spatial metric is calculated as given below

$$
\gamma_{ij} = \begin{pmatrix} \sinh^2 u + \sin^2 v & 0 & 0 \\ 0 & \sinh^2 u + \sin^2 v & 0 \\ 0 & 0 & 1 \end{pmatrix}; \tag{31}
$$

$$
\gamma = \det(\gamma_{ij}) = \sinh^2 u + \sin^2 v \tag{32}
$$

Assuming, the region where the object to be hidden $0 < u < U1$, and the region of invisibility cloaking shell is $U1 < u < U2$, then the physical medium w.r.t. the primed empty curved space-time is given by

(a)

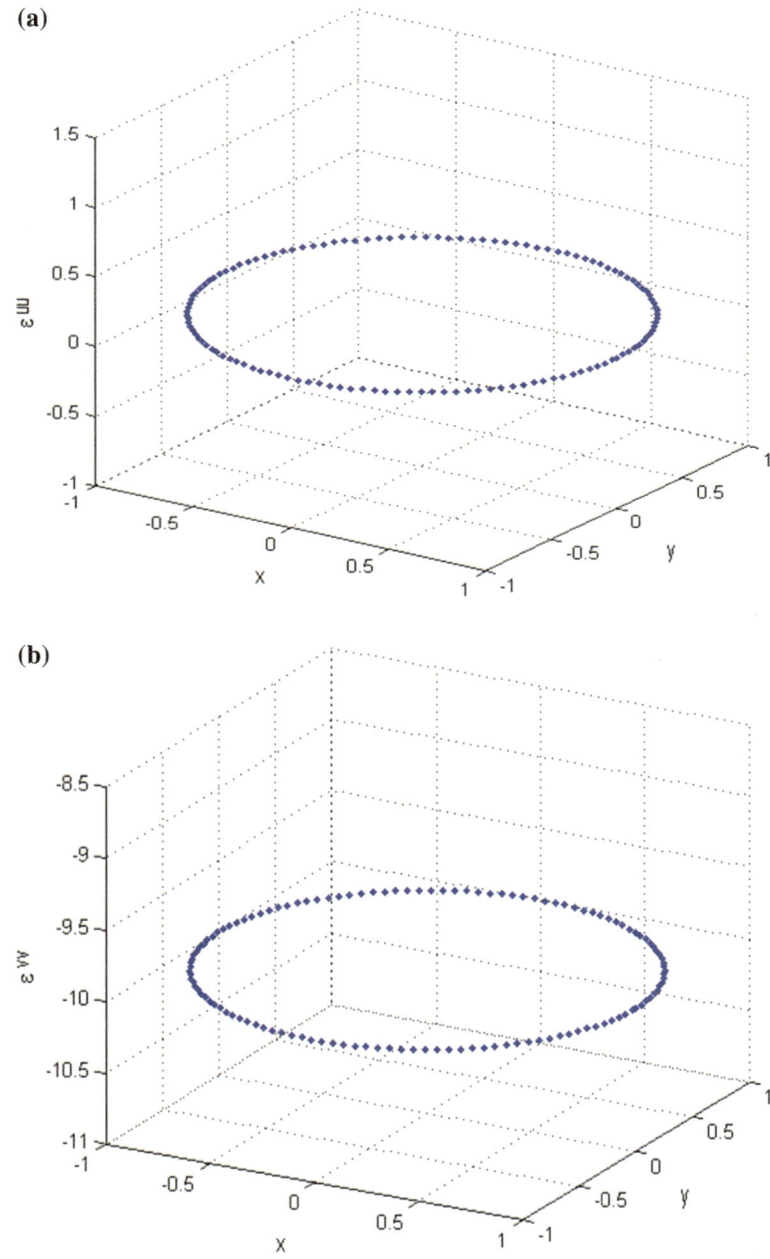

(b)

Fig. 3 **a** Permittivity tensor ε^{uu} distribution inside the right circular cloaking shell. **b** Permittivity tensor ε^{vv} distribution inside the right circular cloaking shell. **c** Permittivity tensor ε^{zz} distribution inside the right circular cloaking shell

(c)

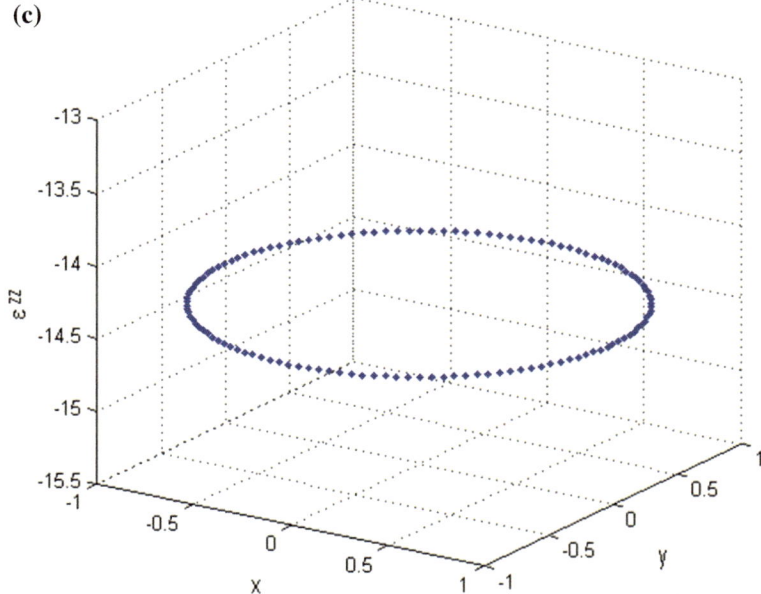

Fig. 3 (continued)

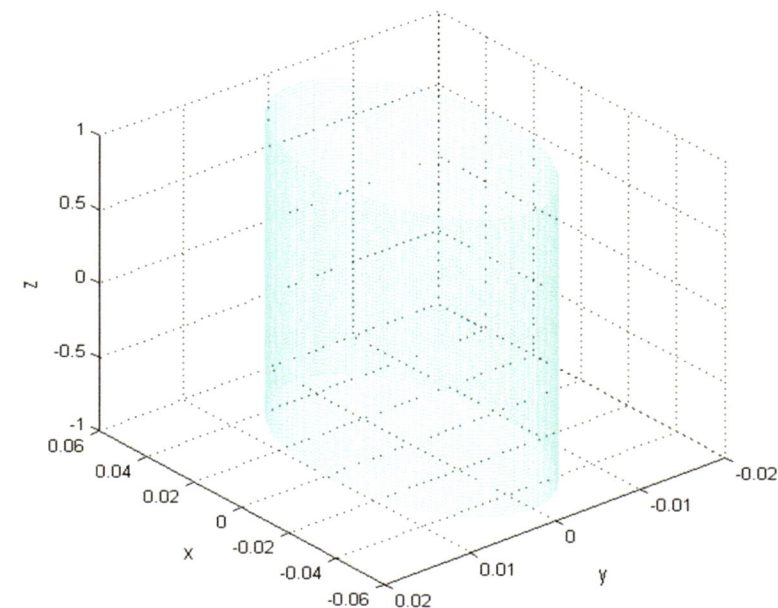

Fig. 4 Boundary of a elliptic cylindrical cloaking shell

$$u = U1 + u' * \frac{(U2 - U1)}{U2}; \quad v = v'; \quad z = z'; \tag{33}$$

$$\partial u /_{\partial u'} = \frac{(U2 - U1)}{U2} \tag{34}$$

The effective geometry corresponding to the bi-anisotropic medium can be defined as

$$g_{ij} = \frac{\partial x^i}{\partial x'^k} \frac{\partial x^j}{\partial x'^l} \gamma'^{kl} = \begin{pmatrix} \left(\frac{U2-U1}{U2}\right)^2 * \frac{1}{\sinh^2 u' + \sin^2 v'} & 0 & 0 \\ 0 & \frac{1}{\sinh^2 u' + \sin^2 v'} & 0 \\ 0 & 0 & 1 \end{pmatrix}; \tag{35}$$

$$\sqrt{-g} = \sqrt{-g'} * \frac{\partial(x'^1, x'^2, x'^3)}{\partial(x^1, x^2, x^3)} = (\sinh^2 u' + \sin^2 v') * \left(\frac{U2}{U2 - U1}\right); \tag{36}$$

$$\sqrt{\gamma} = \sinh^2 u + \sin^2 v \tag{37}$$

The permittivity tensors are given by

$$\varepsilon^{ij} = \pm \frac{\sqrt{-g}}{\sqrt{\gamma}} * g^{ij} \tag{38}$$

Hence the permittivity tensor for a right elliptic cylindrical cloaking shell can be derived using Eqs. (34) through (37) as:

$$\varepsilon^{ij} = \frac{\frac{U2}{U2-U1}}{\sinh^2 u + \sin^2 v} * (\sinh^2 u' + \sin^2 v')$$

$$* \begin{pmatrix} \left(\frac{U2-U1}{U2}\right)^2 * \frac{1}{\sinh^2 u' + \sin^2 v'} & 0 & 0 \\ 0 & \frac{1}{\sinh^2 u' + \sin^2 v'} & 0 \\ 0 & 0 & 1 \end{pmatrix} \tag{39}$$

$$= \begin{pmatrix} \left(\frac{U2-U1}{U2}\right) & 0 & 0 \\ 0 & \frac{U2}{U2-U1} & 0 \\ 0 & 0 & \frac{U2}{U2-U1} * (\sinh^2 u' + \sin^2 v') \end{pmatrix} * \frac{1}{\sinh^2 u + \sin^2 v}$$

For simulation studies a right elliptic cylinder having major axis and minor axis as 0.1, 0.05 respectively is considered. The values of the boundary cloaking shell is taken as, $U1 = 0.1$, $U2 = 0.2$ and the boundary of the object to be cloaked is considered as $u = 0.1$ (for inline cloaking we made u and $U1$ similar). To hide a cylindrical object of dimension $0 < u < U1$, the permittivity tensors in the uu, vv and zz directions are given in Fig. 5a through c.

(a)

(b)

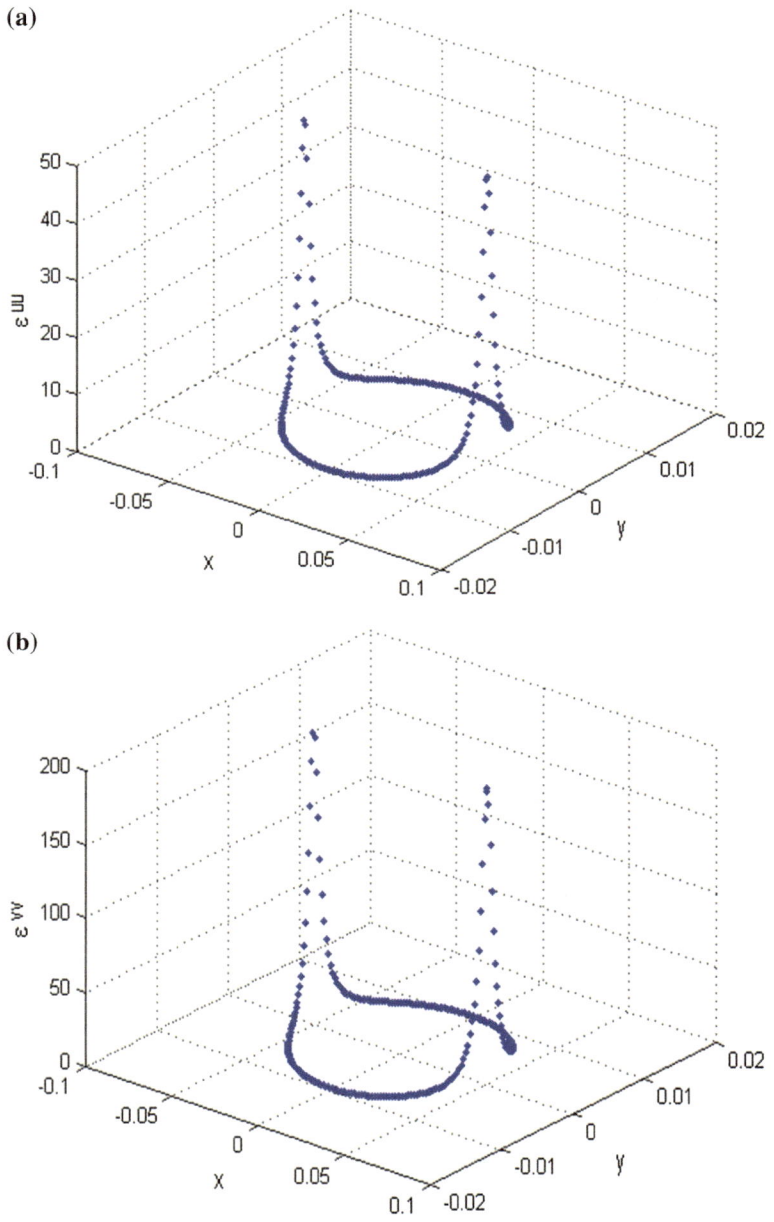

Fig. 5 a Permittivity tensor ε^{uu} distribution inside an elliptic cylindrical cloaking shell. **b** Permittivity tensor ε^{vv} distribution inside an elliptic cylindrical cloaking shell. **c** Permittivity tensor ε^{zz} distribution inside an elliptic cylindrical cloaking shell

(c)

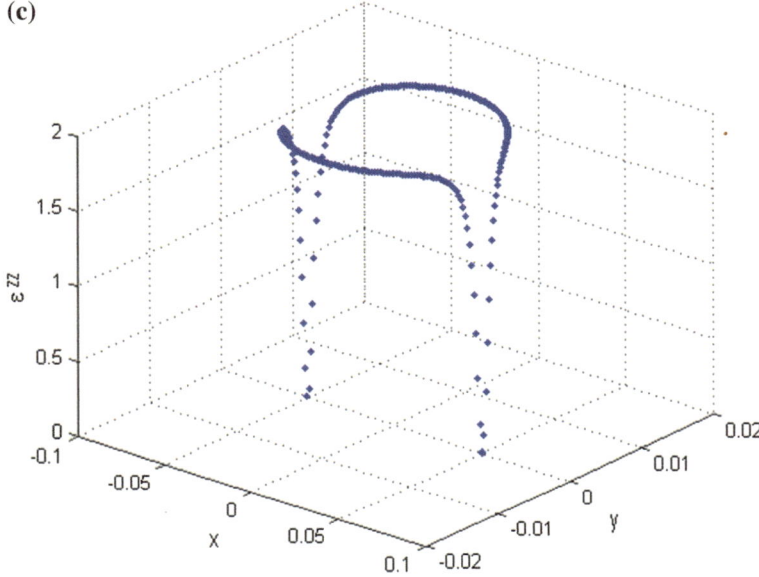

Fig. 5 (continued)

3.3 Permittivity Tensor for Hyperbolic Cylindrical Cloaking Shell

The variation of permittivity tensors for a right hyperbolic cylindrical structure is studied in this section which can be further used for generating hybrid structures. The parametric equation of a right hyperbolic cylinder is given as:

$$x = \cosh u * \cos v; \quad y = \sinh u * \sin v; \quad z = z \tag{40}$$

where u is $(0 \le \infty)$, v is $(-\infty$ to $\infty)$ and z is the finite height of the cylinder. The schematic of a right hyperbolic cylindrical structure is given in Fig. 6.

The position vector of the above mentioned parametric equation was derived (Appendix C) and the spatial metric is calculated as given below

$$\gamma_{ij} = \begin{pmatrix} \sinh^2 u + \sin^2 v & 0 & 0 \\ 0 & \sinh^2 u + \sin^2 v & 0 \\ 0 & 0 & 1 \end{pmatrix}; \tag{41}$$

$$\gamma = \det(\gamma_{ij}) = \sinh^2 u + \sin^2 v \tag{42}$$

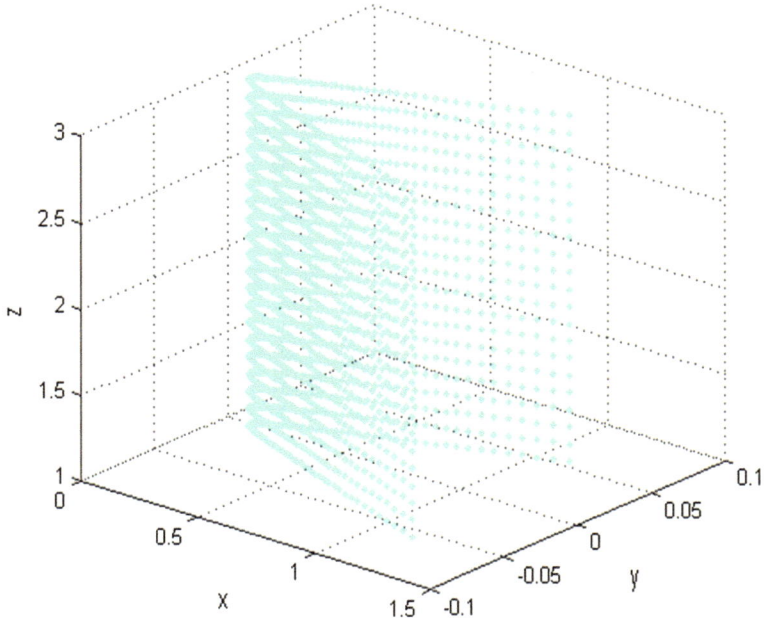

Fig. 6 Boundary of a hyperbolic cylindrical cloaking shell

Assuming, the region where the object to be hidden $0 < u < U1$, and the region of invisibility cloaking shell is $U1 < u < U2$, then the physical medium w.r.t. the primed empty curved space-time is given by

$$u = U1 + u' * \frac{(U2 - U1)}{U2}; \quad v = v'; \quad z = z'; \qquad (43)$$

$$\partial u/\partial u' = \frac{(U2 - U1)}{U2} \qquad (44)$$

The effective geometry corresponding to the bi-anisotropic medium can be defined as

$$g_{ij} = \frac{\partial x^i}{\partial x'^k} \frac{\partial x^j}{\partial x'^l} \gamma'^{kl} = \begin{pmatrix} \left(\frac{U2-U1}{U2}\right)^2 * \frac{1}{\sinh^2 u' + \sin^2 v'} & 0 & 0 \\ 0 & \frac{1}{\sinh^2 u' + \sin^2 v'} & 0 \\ 0 & 0 & 1 \end{pmatrix}; \qquad (45)$$

$$\sqrt{-g} = \sqrt{-g'} * \frac{\partial(x'^1, x'^2, x'^3)}{\partial(x^1, x^2, x^3)} = (\sinh^2 u' + \sin^2 v') * \left(\frac{U2}{U2 - U1}\right); \qquad (46)$$

$$\sqrt{\gamma} = \sinh^2 u + \sin^2 v \qquad (47)$$

The permittivity tensors are given by

$$\varepsilon^{ij} = \pm \frac{\sqrt{-g}}{\sqrt{\gamma}} * g^{ij} \tag{48}$$

Hence the permittivity tensor for a right hyperbolic cylindrical cloaking shell can be derived using Eqs. (45) through (47) as:

$$
\varepsilon^{ij} = \frac{\frac{U2}{U2-U1}}{\sinh^2 u + \sin^2 v} * (\sinh^2 u' + \sin^2 v') * \begin{pmatrix} \left(\frac{U2-U1}{U2}\right)^2 * \frac{1}{\sinh^2 u' + \sin^2 v'} & 0 & 0 \\ 0 & \frac{1}{\sinh^2 u' + \sin^2 v'} & 0 \\ 0 & 0 & 1 \end{pmatrix}
$$

$$
= \begin{pmatrix} \left(\frac{U2-U1}{U2}\right) & 0 & 0 \\ 0 & \frac{U2}{U2-U1} & 0 \\ 0 & 0 & \frac{U2}{U2-U1} * \left(\sinh^2 u' + \sin^2 v'\right) \end{pmatrix} * \frac{1}{\sinh^2 u + \sin^2 v} \tag{49}
$$

permittivity tensor derived is used for simulation of typical right hyperbolic cylinder having dimensions, $u = 0.1$, $U1 = 0.1$, $U2 = 0.2$, height of the cylinder varies from -5 to 5 and v varies from $0°$ to $360°$. To hide a cylindrical object of dimension $0 < u \le U1$, the permittivity tensors in the uu, vv and zz directions are given in Fig. 7a through c.

3.4 Permittivity Tensor for Parabolic Cylindrical Cloaking Shell

The parametric equation of a right parabolic cylinder is given as:

$$x = \frac{1}{2}\left(u^2 - v^2\right); \quad y = uv; \quad z = z \tag{50}$$

where, u is $(0 \le \infty)$, v is $(-\infty$ to $\infty)$ and z is the finite height of the cylinder. Figure 8 shows the right parabolic cylinder generated using Matlab.

The position vector of the above mentioned parametric equation was derived (Appendix D) and the spatial metric is calculated as given below

$$\gamma_{ij} = \begin{pmatrix} u^2 + v^2 & 0 & 0 \\ 0 & u^2 + v^2 & 0 \\ 0 & 0 & 1 \end{pmatrix}; \tag{51}$$

$$\gamma = \det(\gamma_{ij}) = \left(u^2 + v^2\right)^2 \tag{52}$$

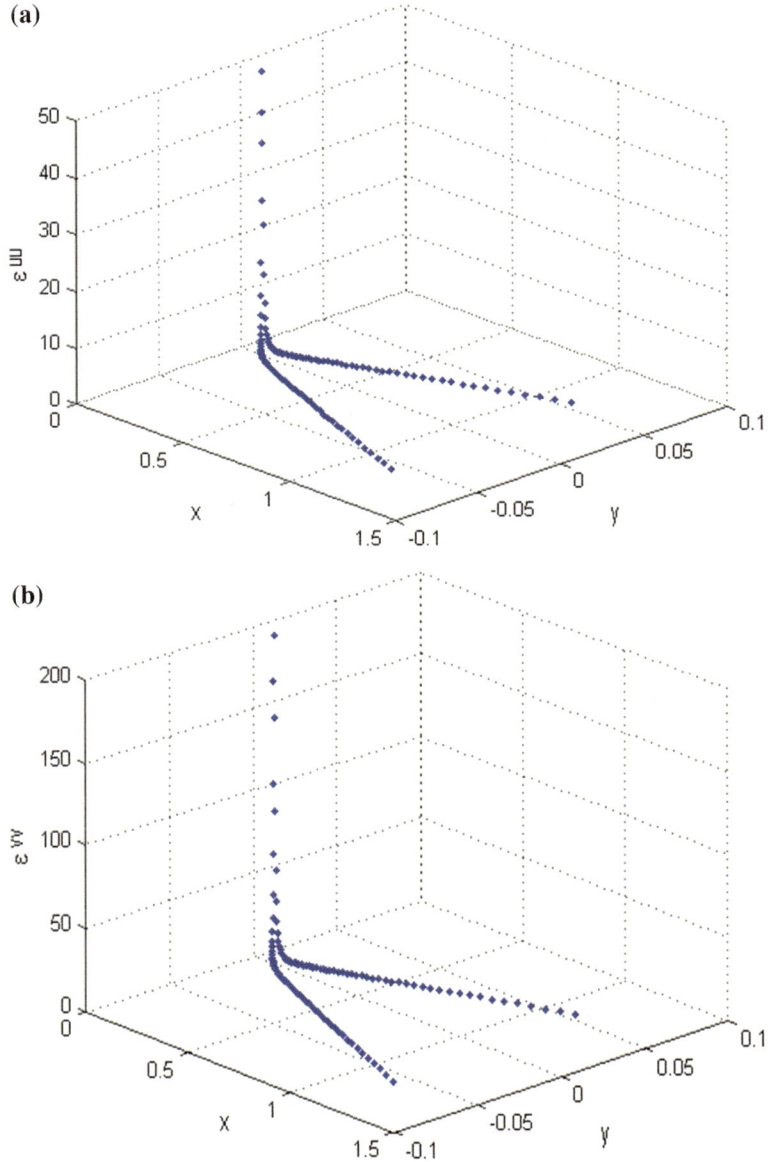

Fig. 7 a Permittivity tensor ε^{uu} distribution inside the hyperbolic cylindrical invisibility device. **b** Permittivity tensor ε^{vv} distribution inside the hyperbolic cylindrical invisibility device. **c** Permittivity tensor ε^{zz} distribution inside the hyperbolic cylindrical invisibility device

(c)

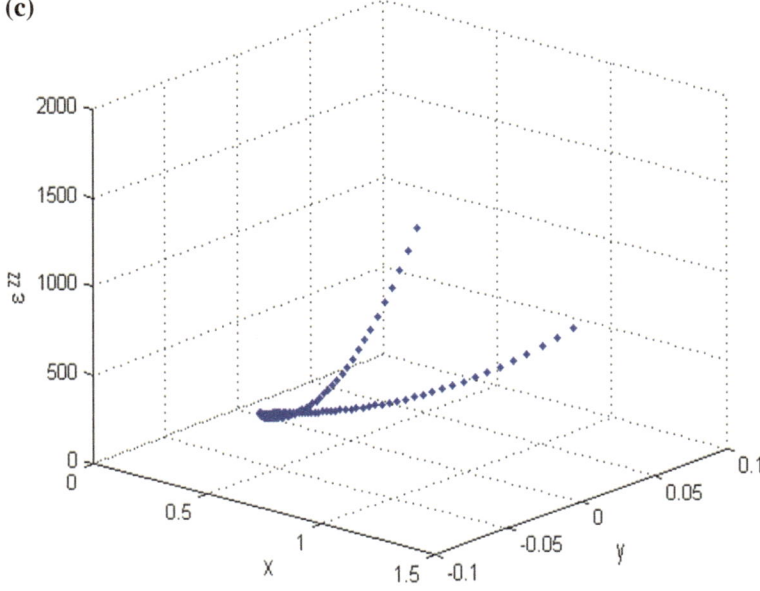

Fig. 7 (continued)

Assuming, the region where the object to be hidden $0 < u < U1$, and the region of invisibility cloaking shell is $U1 < u < U2$, then the physical medium w.r.t. the primed empty curved space-time is given by

$$u = U1 + u' * \frac{(U2 - U1)}{U2}; \quad v = v'; \quad z = z';$$ (53)

$$\partial u/\partial u' = \frac{(U2 - U1)}{U2}$$ (54)

The effective geometry corresponding to the bi-anisotropic medium can be defined as

$$g_{ij} = \frac{\partial x^i}{\partial x'^k} \frac{\partial x^j}{\partial x'^l} \gamma'^{kl} = \begin{pmatrix} \left(\frac{U2-U1}{U2}\right)^2 * \frac{1}{u'^2+v'^2} & 0 & 0 \\ 0 & \frac{1}{u'^2+v'^2} & 0 \\ 0 & 0 & 1 \end{pmatrix};$$ (55)

$$\sqrt{-g} = \sqrt{-g'} * \frac{\partial(x'^1, x'^2, x'^3)}{\partial(x^1, x^2, x^3)} = (u'^2 + v'^2) * \left(\frac{U2}{U2 - U1}\right);$$ (56)

$$\sqrt{\gamma} = u^2 + v^2$$ (57)

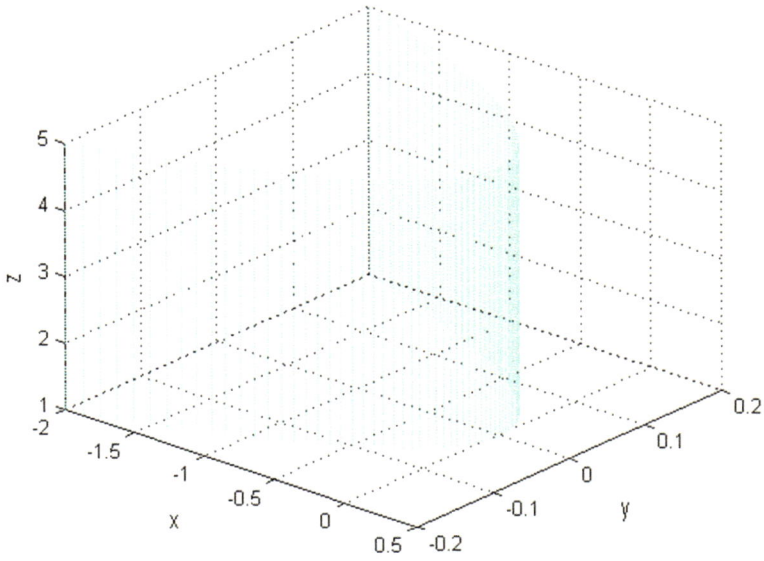

Fig. 8 Boundary of a right parabolic cylindrical cloaking shell

The permittivity tensors are given by

$$\varepsilon^{ij} = \pm \frac{\sqrt{-g}}{\sqrt{\gamma}} * g^{ij} \tag{58}$$

Hence the permittivity tensor for a right parabolic cylindrical cloaking shell can be derived using Eqs. (55) through (57) as:

$$\varepsilon^{ij} = \frac{\left(u'^2 + v'^2\right) * \left(\frac{U2}{U2-U1}\right)}{u^2 + v^2}$$

$$* \begin{pmatrix} \left(\frac{U2-U1}{U2}\right)^2 * \frac{1}{u'^2+v'^2} & 0 & 0 \\ 0 & \frac{1}{u'^2+v'^2} & 0 \\ 0 & 0 & 1 \end{pmatrix} \tag{59}$$

$$= \left(\frac{1}{u^2+v^2}\right) * \begin{pmatrix} \left(\frac{U2-U1}{U2}\right) & 0 & 0 \\ 0 & \left(\frac{U2}{U2-U1}\right) & 0 \\ 0 & 0 & \left(\frac{U2}{U2-U1}\right) * \left(u'^2 + v'^2\right) \end{pmatrix}$$

The corresponding simulation results for permittivity tensors in the *uu*, *vv* and *zz* directions for a parabolic cylindrical surface is shown in Fig. 9a through c. The dimensions considered for simulation are $u = 0.1$, $U1 = 0.1$, and $U2 = 0.2$.

(a)

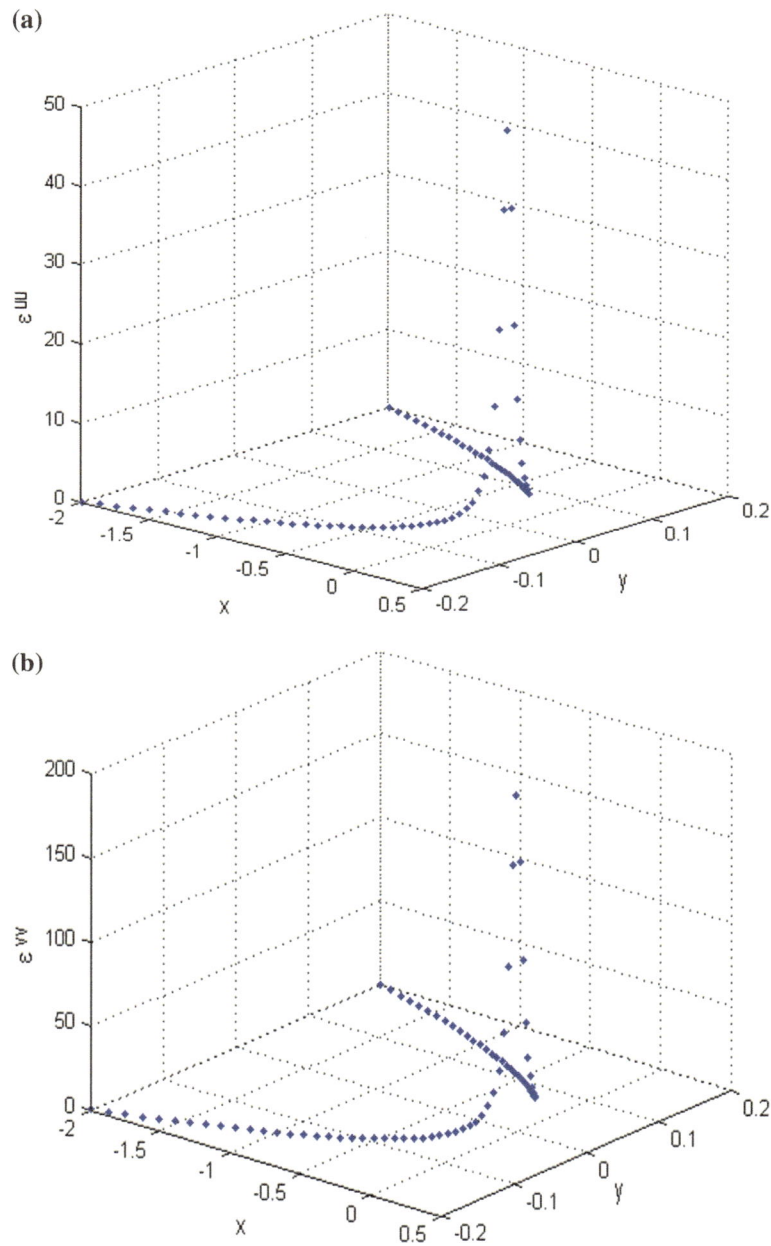

(b)

Fig. 9 **a** Permittivity tensor ε^{uu} distribution inside the parabolic cylindrical invisibility cloaking structure. **b** Permittivity tensor ε^{vv} distribution inside the parabolic cylindrical invisibility cloaking structure. **c** Permittivity tensor ε^{zz} distribution inside parabolic cylindrical invisibility cloaking structure

(c)

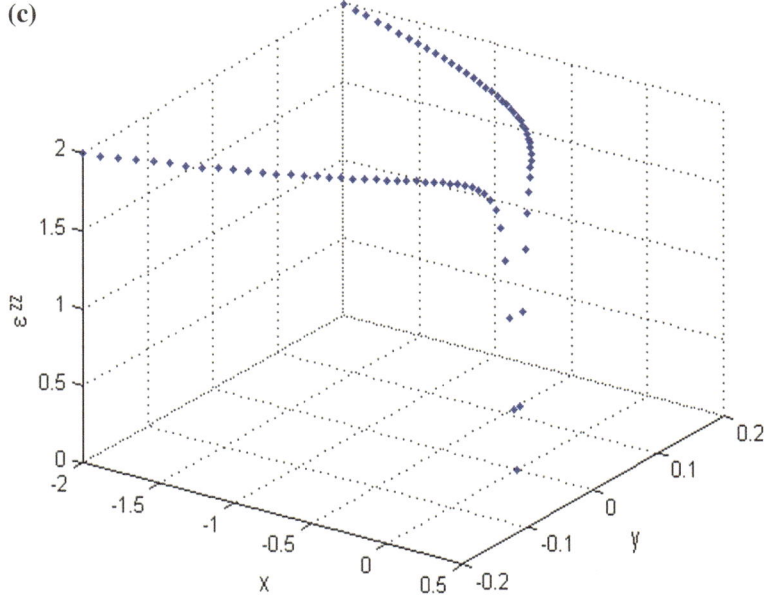

Fig. 9 (continued)

4 Quadric Surfaces of Revolutions

Surface of revolutions are the second degree quadric surfaces that can generate various structures available in nature. Moreover the aerospace structures are made up of hybrids of quadratic cylinders and quadric surface of revolutions. Hence various surfaces of revolutions are studied here towards deriving analytical expressions for cloaking parameters. As the parametric equations are the a priory requirement for calculation of position vectors and spatial metric tensors, the analytical parametric equations of the quadric surfaces of revolutions along with the derivations for permittivity tensors are described in this section. The basic quadric surfaces of revolutions are sphere, cone, prolate spheroid, oblate spheroid etc.

4.1 Permittivity Tensor for Spherical Cloaking Shell

Sphere is a non developable 3D surface that can hide any kind of 3D surfaces. Hence, at first design of spherical cloaking shell is taken up in this section. The parametric equations of a sphere are given as:

$$x = u * \sin v * \cos \varphi; \quad y = u * \sin v * \sin \varphi; \quad z = u * \cos v \qquad (60)$$

where u is the radius of the sphere, v is the azimuth angle varying from $0°$ to $180°$ and φ is the elevation angle that varies from $0°$ to $360°$. Figure 10 shows the boundary of a spherical cloaking shell generated using Matlab.

The position vector of the spherical surface was derived using above mentioned parametric equation (Appendix E) and the spatial metric is calculated as given below (Fig. 12)

$$\gamma_{ij} = \begin{pmatrix} 1 & 0 & 0 \\ 0 & u^2 & 0 \\ 0 & 0 & u^2 * \sin^2 v \end{pmatrix}; \quad \gamma = \det(\gamma_{ij}) = u^4 * \sin^2 v \tag{61}$$

Assuming, the region where the object to be hidden $0 < u < U1$, and the region of cloaking shell is $U1 < u < U2$, the physical medium w.r.t. the primed empty curved space-time is given by

$$u = U1 + u' * \frac{(U2 - U1)}{U2}; \quad v = v'; \quad \varphi = \varphi'; \tag{62}$$

$$\frac{\partial u}{\partial u'} = \frac{(U2 - U1)}{U2} \tag{63}$$

The effective geometry corresponding to the bi-anisotropic medium can be defined as

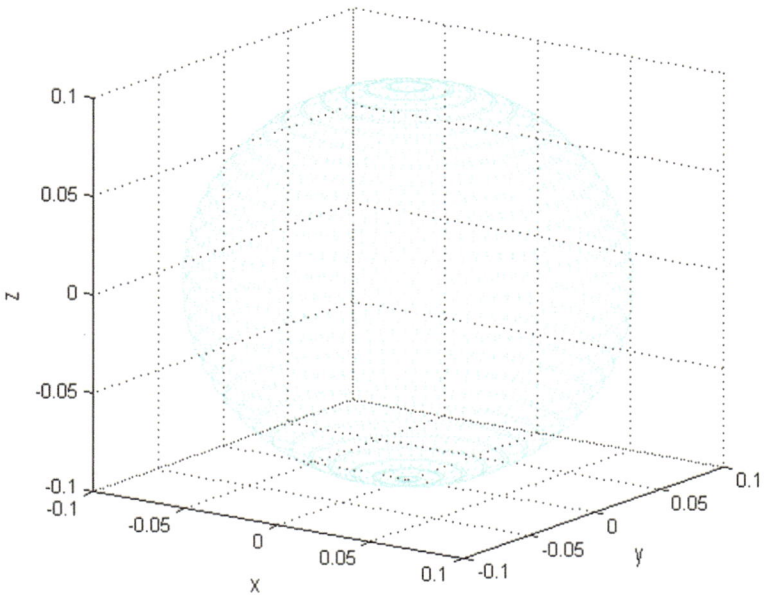

Fig. 10 The boundary of the spherical cloaking shell

$$g^{ij} = \begin{pmatrix} \left(\frac{U2-U1}{U2}\right)^2 & 0 & 0 \\ 0 & \frac{1}{u'^2} & 0 \\ 0 & 0 & \frac{1}{u'^2 \, * \, \sin^2 v'} \end{pmatrix} \tag{64}$$

$$\sqrt{-g} = \sqrt{-g'} * \frac{\partial(x'^1, x'^2, x'^3)}{\partial(x^1, x^2, x^3)} = u'^2 * \sin v' * \left(\frac{U2}{U2-U1}\right) \tag{65}$$

$$\sqrt{\gamma} = u^2 * \sin v \tag{66}$$

Using Eqs. (62)–(64), the permittivity tensors are derived as $\varepsilon^{ij} = \pm \frac{\sqrt{-g}}{\sqrt{\gamma}} * g^{ij}$

$$\varepsilon^{ij} = \frac{u'^2 * \sin v' * \left(\frac{U2}{U2-U1}\right)}{u^2 * \sin v} * \begin{pmatrix} \left(\frac{U2-U1}{U2}\right)^2 & 0 & 0 \\ 0 & \frac{1}{u'^2} & 0 \\ 0 & 0 & \frac{1}{u'^2 \, * \, \sin^2 v'} \end{pmatrix}$$

$$= \begin{pmatrix} \left(\frac{U2-U1}{U2}\right) * u'^2 * \sin v' & 0 & 0 \\ 0 & \frac{U2}{U2-U1} * \sin v' & 0 \\ 0 & 0 & \frac{U2}{U2-U1} * \frac{1}{\sin v'} \end{pmatrix} * \frac{1}{u^2 * \sin v} \tag{67}$$

The permittivity tensor derived is used for simulation of typical sphere of radius, $u = 0.1$. The inner and outer radius of the spherical cloaking shell is considered as $U1 = 0.1$, $U2 = 0.2$. To hide a object of dimension $0 < u \leq U1$, the permittivity tensors in the uu, vv and zz directions are given in Fig. 11a through c.

4.2 Permittivity Tensor for Conical Cloaking Shell

A right circular cone can be generated keeping the azimuth angle v of the sphere constant. The conical shapes can be designed as the radome of various aircraft systems. The parametric equation of a cone is given as

$$x = u * \sin v * \cos \varphi; \quad y = u * \sin v * \sin \varphi; \quad z = u * \cos v \tag{68}$$

where u is the radius of the sphere from which the right circular cone is generated, v is a constant angle lies between $0°$ and $180°$ and φ is the elevation angle varying from $0°$ to $360°$.

The position vector of the conical surface was derived using the above mentioned parametric equation (Appendix F) and the spatial metric is calculated as given below

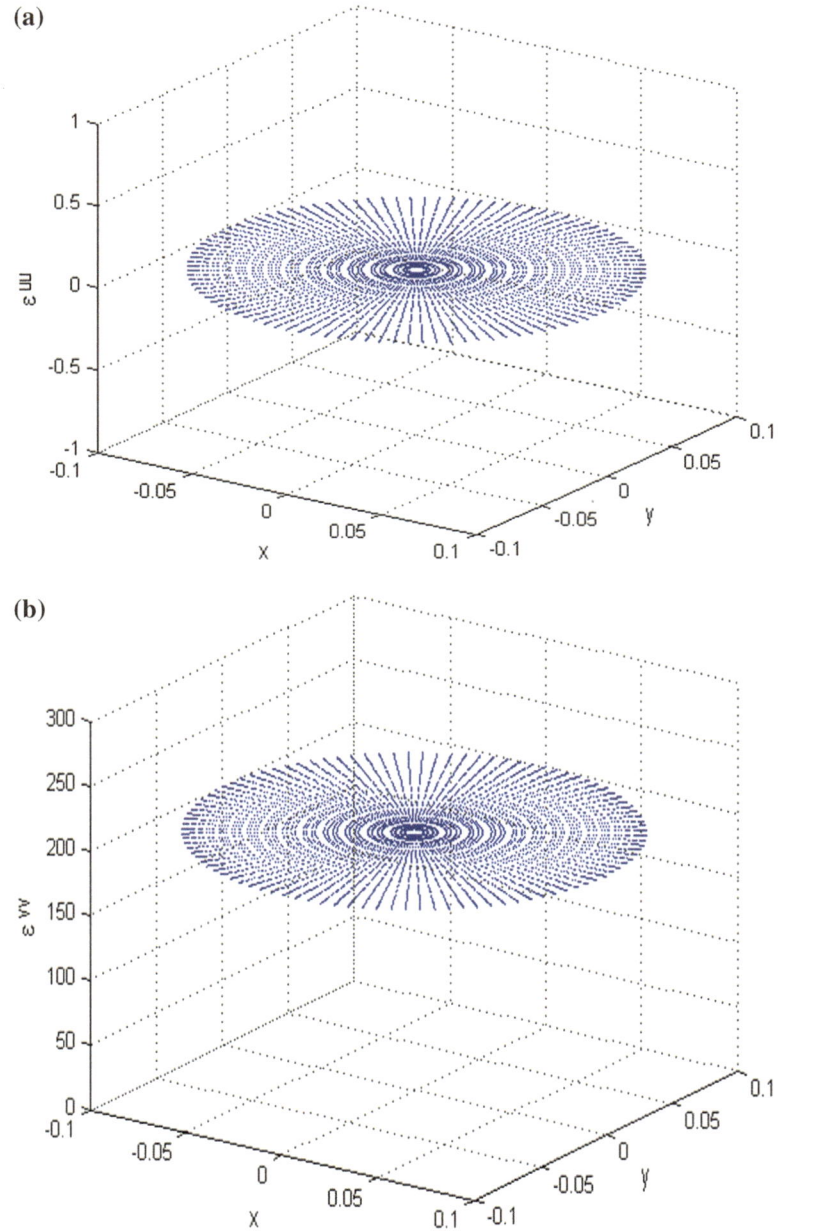

Fig. 11 **a** Permittivity tensor ε^{uu} distribution inside the spherical cloaking shell, **b** Permittivity tensor ε^{vv} distribution inside the spherical cloaking shell, **c** Permittivity tensor ε^{zz} distribution inside the spherical cloaking shell

(c)

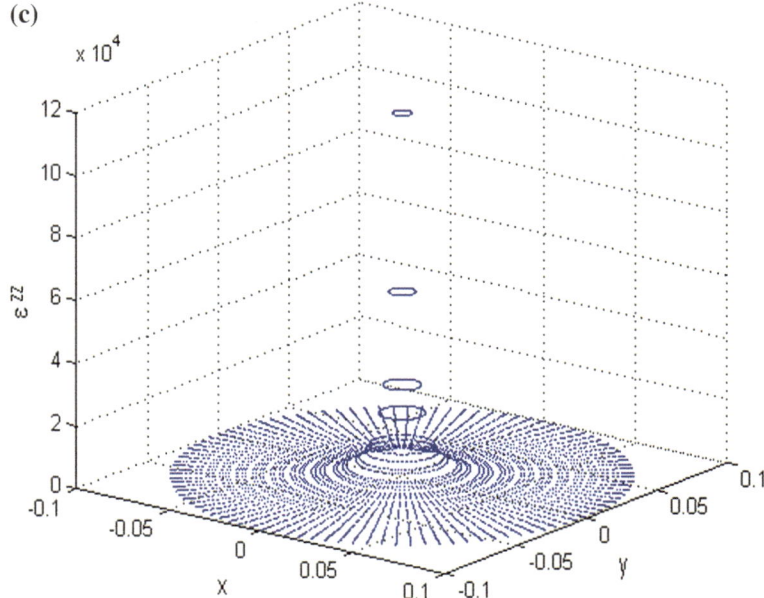

Fig. 11 (continued)

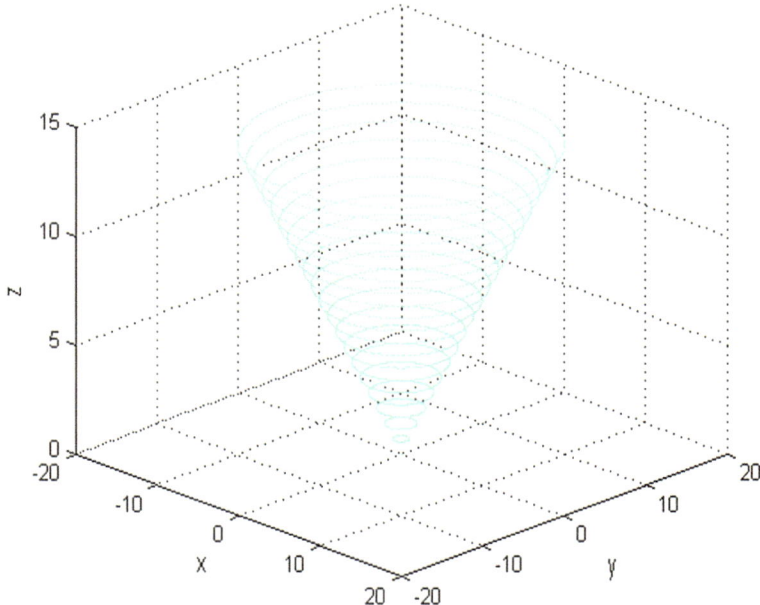

Fig. 12 Boundary of the conical cloaking shell generated using Matlab

$$\gamma_{ij} = \begin{pmatrix} 1 & 0 & 0 \\ 0 & u^2 & 0 \\ 0 & 0 & u^2 * \sin^2 v \end{pmatrix}; \quad \gamma = \det(\gamma_{ij}) = u^4 * \sin^2 v \tag{69}$$

Assuming, the region where the object to be hidden $0 < u < U1$, and the region of cloaking shell is $U1 < u < U2$, the physical medium w.r.t. the primed empty curved space-time is given by

$$u = U1 + u' * \frac{(U2 - U1)}{U2}; \quad v = v'; \quad \varphi = \varphi'; \tag{70}$$

$$\partial u \big/ \partial u' = \frac{(U2 - U1)}{U2} \tag{71}$$

The effective geometry corresponding to the bi-anisotropic medium can be defined as

$$g_{ij} = \frac{\partial x^i}{\partial x'^k} \frac{\partial x^j}{\partial x'^l} \gamma'^{kl} = \begin{pmatrix} \left(\frac{U2-U1}{U2}\right)^2 & 0 & 0 \\ 0 & \frac{1}{u'^2} & 0 \\ 0 & 0 & \frac{1}{u'^2 * \sin^2 v'} \end{pmatrix}; \tag{72}$$

$$\sqrt{-g} = \sqrt{-g'} * \frac{\partial(x'^1, x'^2, x'^3)}{\partial(x^1, x^2, x^3)} = u'^2 * \sin v' * \left(\frac{U2}{U2 - U1}\right); \tag{73}$$

$$\sqrt{\gamma} = u^2 * \sin v \tag{74}$$

The permittivity tensors are given by

$$\varepsilon^{ij} = \pm \frac{\sqrt{-g}}{\sqrt{\gamma}} * g^{ij} \tag{75}$$

Using Eqs. (72)–(74), the permittivity tensors are derived as:

$$\varepsilon^{ij} = \frac{u'^2 * \sin v' * \left(\frac{U2}{U2-U1}\right)}{u^2 * \sin v} * \begin{pmatrix} \left(\frac{U2-U1}{U2}\right)^2 & 0 & 0 \\ 0 & \frac{1}{u'^2} & 0 \\ 0 & 0 & \frac{1}{u'^2 * \sin^2 v'} \end{pmatrix}$$

$$= \begin{pmatrix} \left(\frac{U2-U1}{U2}\right) * u'^2 * \sin v' & 0 & 0 \\ 0 & \frac{U2}{U2-U1} * \sin v' & 0 \\ 0 & 0 & \frac{U2}{U2-U1} * \frac{1}{\sin v'} \end{pmatrix} * \frac{1}{u^2 * \sin v} \tag{76}$$

The permittivity tensor derived (Eq. 76) is used for simulation of typical cone having dimensions $v = 45°$. The conical cloaking shell dimension is considered as

$U1 = 45°$ and $U2 = 47°$. The permittivity tensor distribution inside the cloaking shell in the uu, vv and zz directions are given in Fig. 13a through c.

4.3   Permittivity Tensor for Prolate Spheroidical Cloaking Shell

The parametric equations of a prolate spheroid is given as

$$x = \sinh u * \sin v * \cos \varphi; \quad y = \sinh u * \sin v * \sin \varphi; \quad z = \cosh u * \cos v \quad (77)$$

where u is having range of $0 \leq u \leq \infty$, v is the angle that varies from $0°$ to $180°$ and φ is the angle which varies from $0°$ to $360°$. Figure 14 shows the prolate spheroid generated using Matlab.

The position vector of the above mentioned parametric equation was derived (Appendix G) and the spatial metric is calculated as given below

$$\gamma_{ij} = \begin{pmatrix} \sinh^2 u + \sin^2 v & 0 & 0 \\ 0 & \sinh^2 u + \sin^2 v & 0 \\ 0 & 0 & \sinh^2 u * \sin^2 v \end{pmatrix}; \quad (78)$$

$$\gamma = \det(\gamma_{ij}) = \left((\sinh^2 u + \sin^2 v) * \sinh u * \sin v \right)^2 \quad (79)$$

Assuming, the region where the object to be hidden $0 < u < U1$, and the region of invisibility cloaking shell is $U1 < u < U2$, then the physical medium w.r.t. the primed empty curved space-time is given by

$$u = U1 + u' * \frac{(U2 - U1)}{U2}; \quad v = v'; \quad \varphi = \varphi'; \quad (80)$$

$$\frac{\partial u}{\partial u'} = \frac{(U2 - U1)}{U2}$$

The effective geometry corresponding to the bi-anisotropic medium can be defined as

$$g_{ij} = \frac{\partial x^i}{\partial x'^k} \frac{\partial x^j}{\partial x'^l} \gamma'^{kl} = \begin{pmatrix} \left(\frac{U2-U1}{U2}\right)^2 * \frac{1}{\sinh^2 u' + \sin^2 v'} & 0 & 0 \\ 0 & \frac{1}{\sinh^2 u' + \sin^2 v'} & 0 \\ 0 & 0 & \frac{1}{\sinh^2 u' * \sin^2 v'} \end{pmatrix};$$

$$(81)$$

(a)

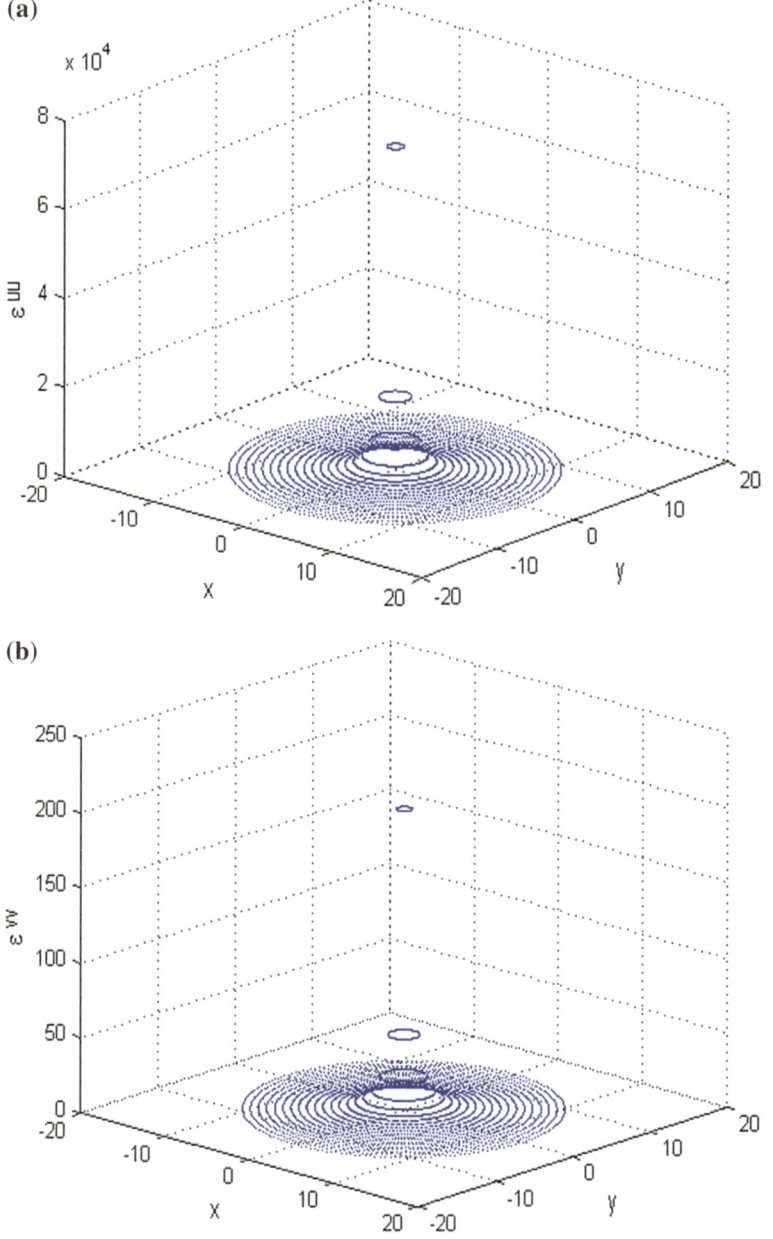

(b)

Fig. 13 **a** Permittivity tensor ε^{uu} distribution inside the invisibility cloak of conical shape.
b Permittivity tensor ε^{vv} distribution inside the invisibility cloaking shell of conical shape.
c Permittivity tensor ε^{zz} distribution inside the invisibility cloaking shell of conical shape

(c)

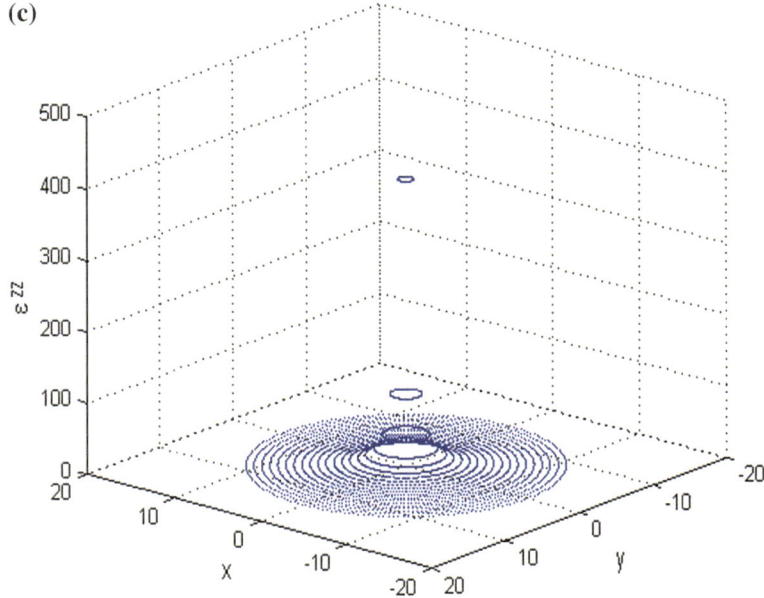

Fig. 13 (continued)

Fig. 14 The prolate
spheroidal structure generated
using Matlab

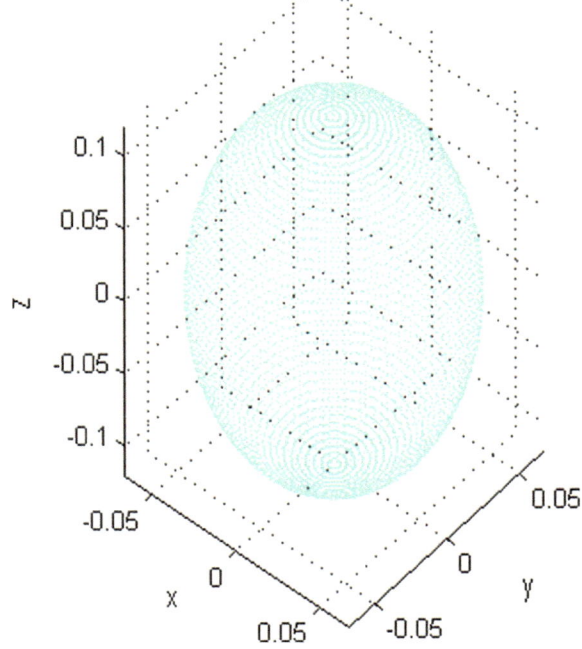

$$\sqrt{-g} = \sqrt{-g'} * \frac{\partial(x'^1, x'^2, x'^3)}{\partial(x^1, x^2, x^3)}$$

$$= (\sinh^2 u' + \sin^2 v') * \sinh u' * \sin v' * \left(\frac{U2}{U2 - U1}\right); \tag{82}$$

$$\sqrt{\gamma} = (\sinh^2 u + \sin^2 v) * \sinh u * \sin v \tag{83}$$

The permittivity tensors are given by

$$\varepsilon^{ij} = \pm \frac{\sqrt{-g}}{\sqrt{\gamma}} * g^{ij} \tag{84}$$

The permittivity tensors can be derived using Eqs. (79) through (83) as:

$$\varepsilon^{ij} = \frac{(\sinh^2 u' + \sin^2 v') * \sinh u' * \sin v' * \left(\frac{U2}{U2-U1}\right)}{(\sinh^2 u + \sin^2 v) * \sinh u * \sin v}$$

$$* \begin{pmatrix} \left(\frac{U2-U1}{U2}\right)^2 * \frac{1}{\sinh^2 u' + \sin^2 v'} & 0 & 0 \\ 0 & \frac{1}{\sinh^2 u' + \sin^2 v'} & 0 \\ 0 & 0 & \frac{1}{\sinh^2 u' * \sin^2 v'} \end{pmatrix}$$

$$= \begin{pmatrix} \left(\frac{U2-U1}{U2}\right) * \sinh u' * \sin v' & 0 & 0 \\ 0 & \left(\frac{U2}{U2-U1}\right) * \sinh u' * \sin v' & 0 \\ 0 & 0 & \frac{U2}{U2-U1} * \frac{(\sinh^2 u' + \sin^2 v')}{\sinh u' * \sin v'} \end{pmatrix}$$

$$* \frac{1}{(\sinh^2 u + \sin^2 v) * \sinh u * \sin v}$$

$$\tag{85}$$

The permittivity tensor derived is used for simulation of typical cloaking shell of prolate spheroidical shape having dimensions as $u = 0.1$, $U1 = 0.1$, $U2 = 0.2$. To hide a object of dimension $0 < u < U1$, the permittivity tensors in the uu, vv and zz directions are given in Fig. 15a through c.

4.4 Permittivity Tensor for Oblate Spheroidical Cloaking Shell

An invisibility device of oblate spheroidical shape is considered in this section. The parametric equation of an oblate spheroid is given as:

$$x = \cosh u * \cos v * \cos \varphi; \quad y = \cosh u * \cos v * \sin \varphi; \quad z = \sinh u * \sin v \tag{86}$$

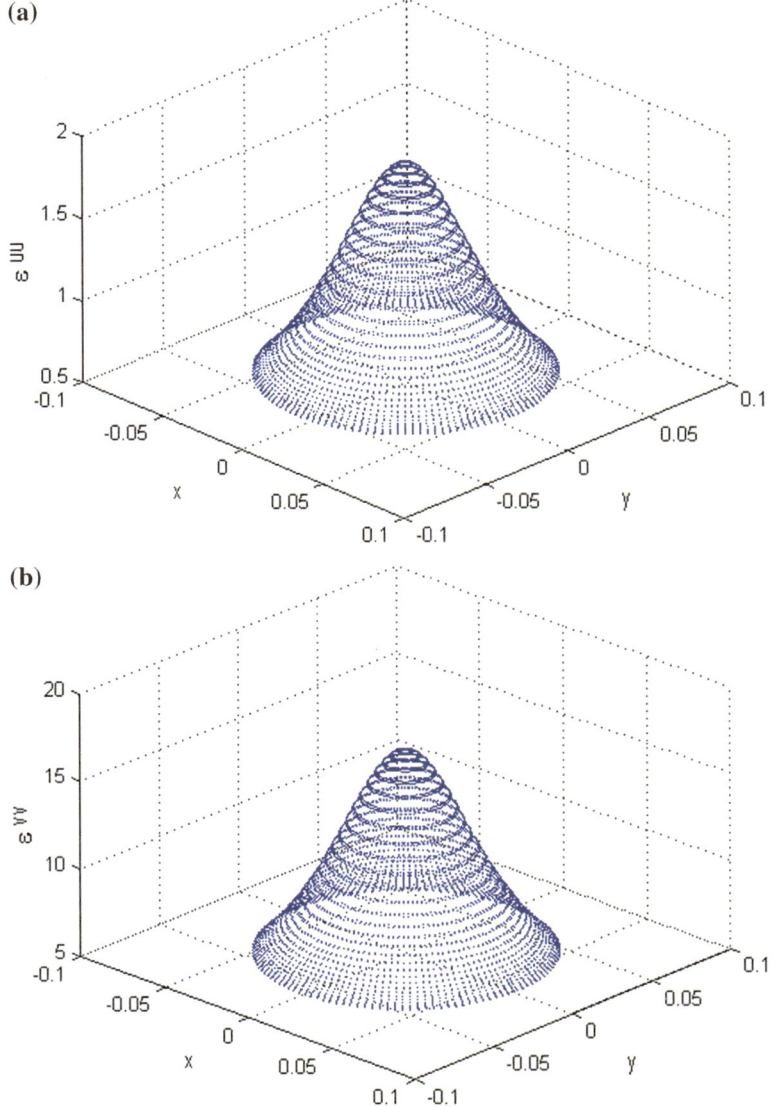

Fig. 15 a Permittivity tensor ε^{uu} distribution inside the invisibility cloaking shell of prolate spheroidical shape. **b** Permittivity tensor ε^{vv} distribution inside the invisibility cloaking shell of prolate spheroidical shape. **c** Permittivity tensor ε^{zz} distribution inside the invisibility cloaking shell of prolate spheriodical shape

where u is a variable which varies from zero to infinity, v is the angle varying from $0°$ to $180°$ and φ is the angle that varies from $0°$ to $360°$. Figure 16 shows the boundary of an oblate spheroid generated using Matlab.

(c)

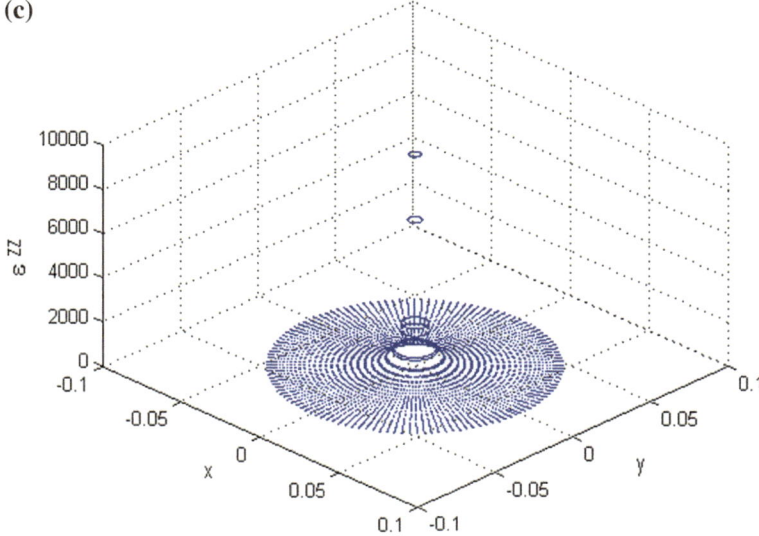

Fig. 15 (continued)

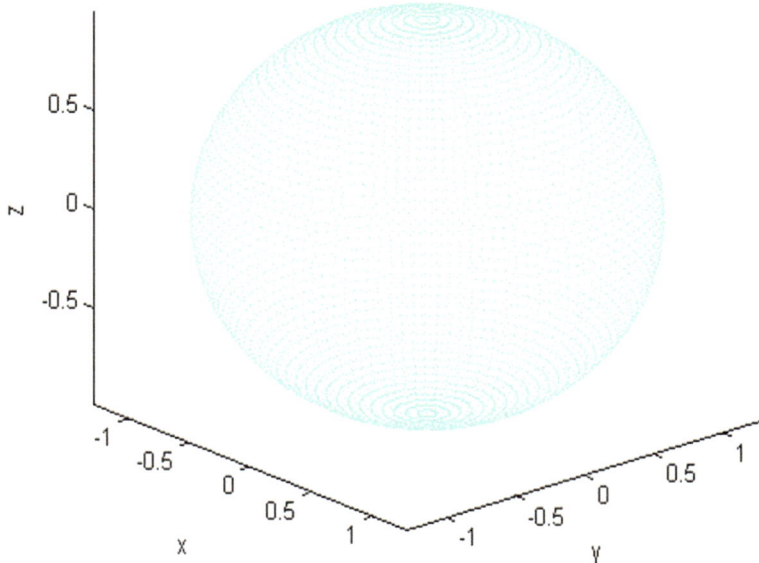

Fig. 16 Boundary of an oblate spheroid

The position vector of the above mentioned parametric equation was derived (Appendix H) and the spatial metric is calculated as given below

$$\gamma_{ij} = \begin{pmatrix} \sinh^2 u + \sin^2 v & 0 & 0 \\ 0 & \sinh^2 u + \sin^2 v & 0 \\ 0 & 0 & \cosh^2 u * \cos^2 v \end{pmatrix}; \qquad (87)$$

$$\gamma = \det(\gamma_{ij}) = \left(\left(\sinh^2 u + \sin^2 v \right) * \sinh u * \sin v \right)^2 \qquad (88)$$

Assuming, the region where the object to be hidden $0 < u < U1$, and the region of invisibility cloaking shell is $U1 < u < U2$, the physical medium w.r.t. the primed empty curved space-time is given by

$$u = U1 + u' * \frac{(U2 - U1)}{U2}; \quad v = v'; \quad \varphi = \varphi'; \qquad (89)$$

$$\partial u/_{\partial u'} = \frac{(U2 - U1)}{U2} \qquad (90)$$

The effective geometry corresponding to the bi-anisotropic medium can be defined as

$$g_{ij} = \frac{\partial x^i}{\partial x'^k} \frac{\partial x^j}{\partial x'^l} \gamma'^{kl} = \begin{pmatrix} \left(\frac{U2-U1}{U2} \right)^2 * \frac{1}{\sinh^2 u' + \sin^2 v'} & 0 & 0 \\ 0 & \frac{1}{\sinh^2 u' + \sin^2 v'} & 0 \\ 0 & 0 & \frac{1}{\cosh^2 u' * \cos^2 v'} \end{pmatrix}; \qquad (91)$$

$$\sqrt{-g} = \sqrt{-g'} * \frac{\partial(x'^1, x'^2, x'^3)}{\partial(x^1, x^2, x^3)}$$
$$= (\sinh^2 u' + \sin^2 v') * \cosh u' * \cos v' * \left(\frac{U2}{U2 - U1} \right), \qquad (92)$$

$$\sqrt{\gamma} = \left(\sinh^2 u + \sin^2 v \right) * \cosh u * \cos v \qquad (93)$$

Hence, the permittivity tensor can be derived using Eqs. (88) through (91) as:

$$\varepsilon^{ij} = \pm \frac{\sqrt{-g}}{\sqrt{\gamma}} * g^{ij}$$

$$\varepsilon^{ij} = \frac{(\sinh^2 u' + \sin^2 v') * \cosh u' * \cos v' * \left(\frac{U2}{U2-U1}\right)}{(\sinh^2 u + \sin^2 v) * \cosh u * \cos v}$$

$$* \begin{pmatrix} \left(\frac{U2-U1}{U2}\right)^2 * \frac{1}{\sinh^2 u' + \sin^2 v'} & 0 & 0 \\ 0 & \frac{1}{\sinh^2 u' + \sin^2 v'} & 0 \\ 0 & 0 & \frac{1}{\cosh^2 u' * \cos^2 v'} \end{pmatrix}$$

$$= \begin{pmatrix} \left(\frac{U2-U1}{U2}\right) * \cosh u' * \cos v' & 0 & 0 \\ 0 & \left(\frac{U2}{U2-U1}\right) * \cosh u' * \cos v' & 0 \\ 0 & 0 & \frac{U2}{U2-U1} * \frac{(\sinh^2 u' + \sin^2 v')}{\cosh u' * \cos v'} \end{pmatrix}$$

$$* \frac{1}{(\sinh^2 u + \sin^2 v) * \cosh u * \cos v}$$

$$\tag{94}$$

The permittivity tensor derived is used for simulation of typical oblate sphe-roidical cloaking shell of dimensions $u = 0.1, U1 = 0.1$, $U2 = 0.2$. The permittivity tensors in the uu, vv and zz directions are given in Fig. 17a through c.

4.5 Permittivity Tensor for Cloaking Shell of General Paraboloid of Revolution (GPOR) Shape

GPOR is the most important structure in aerospace domain because aerospace structures can be modeled using the hybrids of GPOR. The important aerospace EM structure is the radome which also can be modeled as a GPOR. The parametric equations of a general paraboloid of revolution is given as

$$x = uv \cos \varphi; \quad y = uv \sin \varphi; \quad z = \frac{1}{2}\left(u^2 - v^2\right) \tag{95}$$

where u is the distance between the vertex and focus which varies from zero to infinity v varies from zero to infinity and φ is the angle varying from $0°$ to $360°$. Figure 18 shows the GPOR generated using Matlab.

The position vector of the above mentioned parametric equation was derived (Appendix I) and the spatial metric is calculated as given in Eq. (96).

$$\gamma_{ij} = \begin{pmatrix} u^2 + v^2 & 0 & 0 \\ 0 & u^2 + v^2 & 0 \\ 0 & 0 & u^2 v^2 \end{pmatrix}; \tag{96}$$

$$\gamma = \det(\gamma_{ij}) = u^2 v^2 \left(u^2 + v^2\right)^2; \tag{97}$$

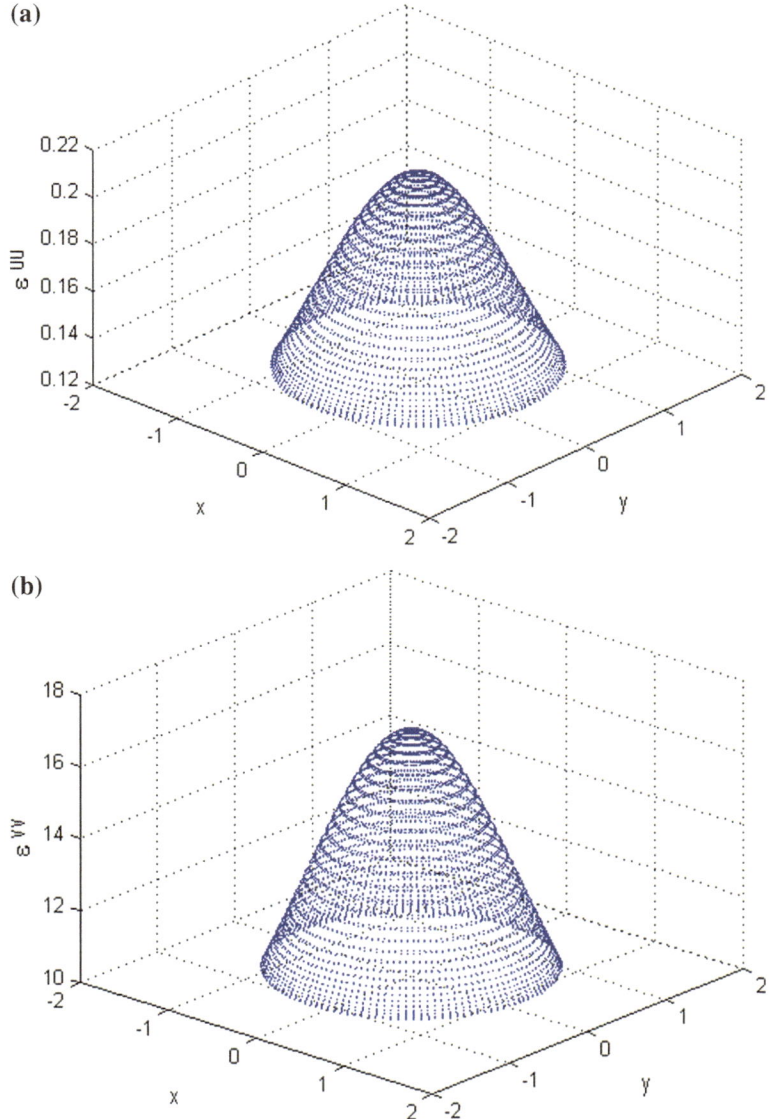

Fig. 17 **a** Permittivity tensor ε^{uu} distribution inside the invisibility cloaking shell of oblate spheriodical shape. **b** Permittivity tensor ε^{vv} distribution inside the invisibility cloaking shell of oblate spheriodical shape. **c** Permittivity tensor $\varepsilon^{\varphi\varphi}$ distribution inside the invisibility cloaking shell of oblate spheriodical shape

(c)

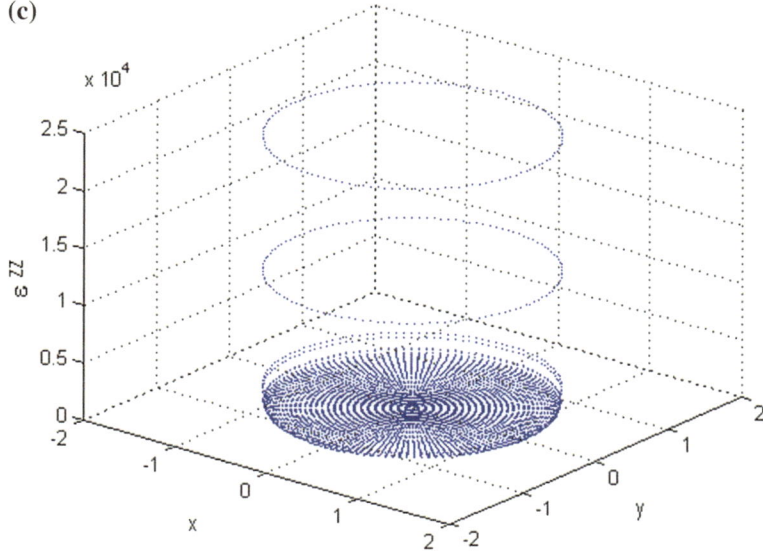

Fig. 17 (continued)

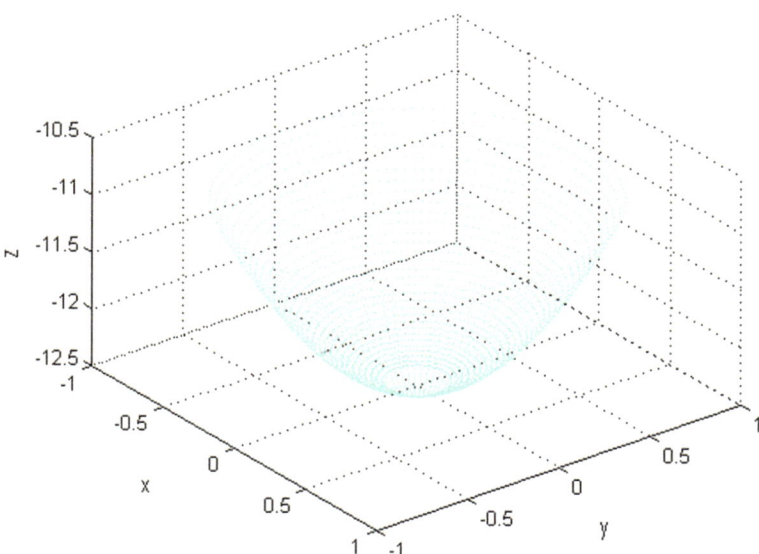

Fig. 18 The boundary of the GPOR in where the object is put into for hiding

Assuming, the region where the object to be hidden $0 < u < U1$, and the region of invisibility cloaking shell is $U1 < u < U2$, then the physical medium w.r.t. the primed empty curved space-time is given by

$$u = U1 + u' * \frac{(U2 - U1)}{U2}; \quad v = v'; \quad \varphi = \varphi'; \tag{98}$$

$$\partial u / \partial u' = \frac{(U2 - U1)}{U2} \tag{99}$$

The effective geometry corresponding to the bi-anisotropic medium can be defined as

$$g_{ij} = \frac{\partial x^i}{\partial x'^k} \frac{\partial x^j}{\partial x'^l} \gamma'^{kl} = \begin{pmatrix} \left(\frac{U_2 - U_1}{U_2}\right)^2 \frac{1}{u'^2 + v'^2} & 0 & 0 \\ 0 & \frac{1}{u'^2 + v'^2} & 0 \\ 0 & 0 & \frac{1}{u'^2 v'^2} \end{pmatrix}; \tag{100}$$

$$\sqrt{-g} = \sqrt{-g'} * \frac{\partial(x'^1, x'^2, x'^3)}{\partial(x^1, x^2, x^3)} = \frac{U2}{U_2 - U_1} u'v'\left(u'^2 + v'^2\right); \tag{101}$$

$$\sqrt{\gamma} = uv\left(u^2 + v^2\right) \tag{102}$$

Hence, the permittivity tensor can be derived using Eqs. (98) through (100) as:

$$\varepsilon^{ij} = \pm \frac{\sqrt{-g}}{\sqrt{\gamma}} * g^{ij} \tag{103}$$

$$\varepsilon^{ij} = \frac{(u'^2 + v'^2) * u' * v' * \left(\frac{U2}{U2 - U1}\right)}{(u^2 + v^2) * u * v}$$

$$* \begin{pmatrix} \left(\frac{U2 - U1}{U2}\right)^2 * \frac{1}{u'^2 + v'^2} & 0 & 0 \\ 0 & \frac{1}{u'^2 + v'^2} & 0 \\ 0 & 0 & \frac{1}{u'^2 * v'^2} \end{pmatrix}$$

$$\varepsilon^{ij} = \begin{pmatrix} \left(\frac{U2 - U1}{U2}\right) * u' *' & 0 & 0 \\ 0 & \left(\frac{U2}{U2 - U1}\right) * u' * v' & 0 \\ 0 & 0 & \frac{U2}{U2 - U1} * \frac{(u'^2 + v'^2)}{u' * v'} \end{pmatrix}$$

$$* \frac{1}{(u^2 + v^2) * u * v} \tag{104}$$

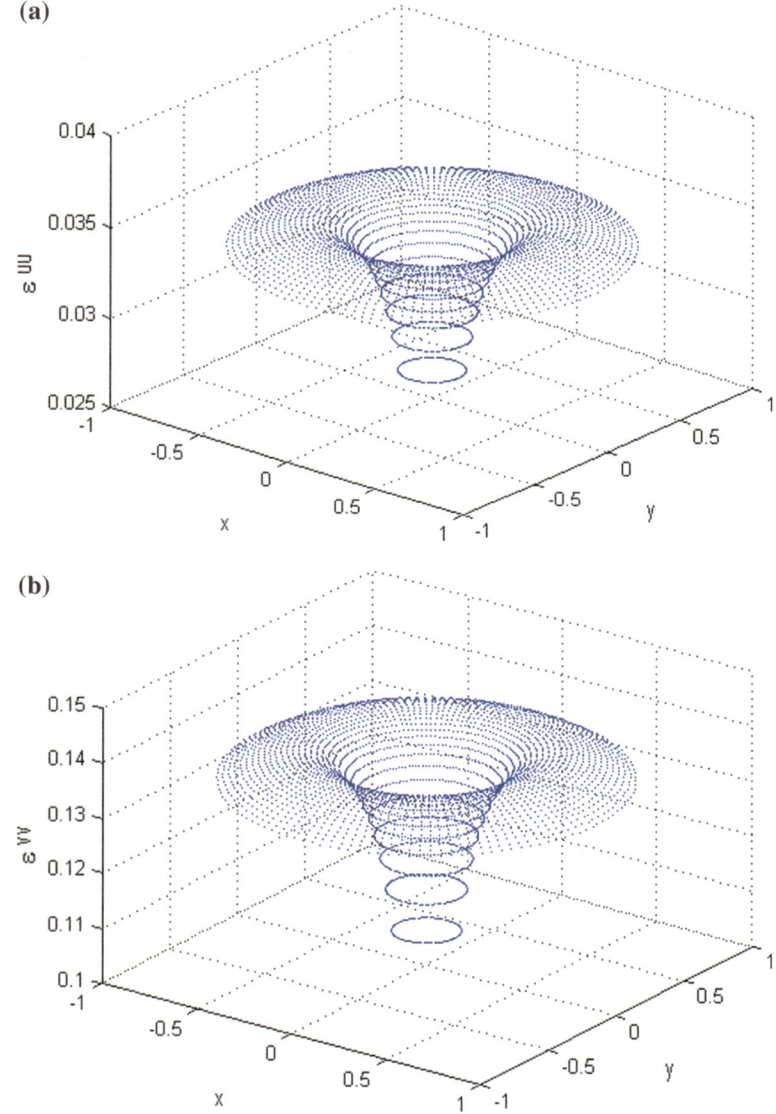

Fig. 19 a Permittivity tensor ε^{uu} distribution inside the invisibility device of GPOR shape. **b** Permittivity tensor ε^{vv} distribution inside the invisibility device of GPOR shape. **c** Permittivity tensor $\varepsilon^{\varphi\varphi}$ distribution inside the invisibility device of GPOR shape

The permittivity tensor derived is used for simulation of typical GPOR having dimensions $u = 0.1$, $U1 = 0.1$, $U2 = 0.2$ and v as $0 \leq v \leq 4$. To hide a object of dimension $0 < u < U1$, the permittivity tensors in the uu, vv and zz directions are given in Fig. 19a through c.

(c)

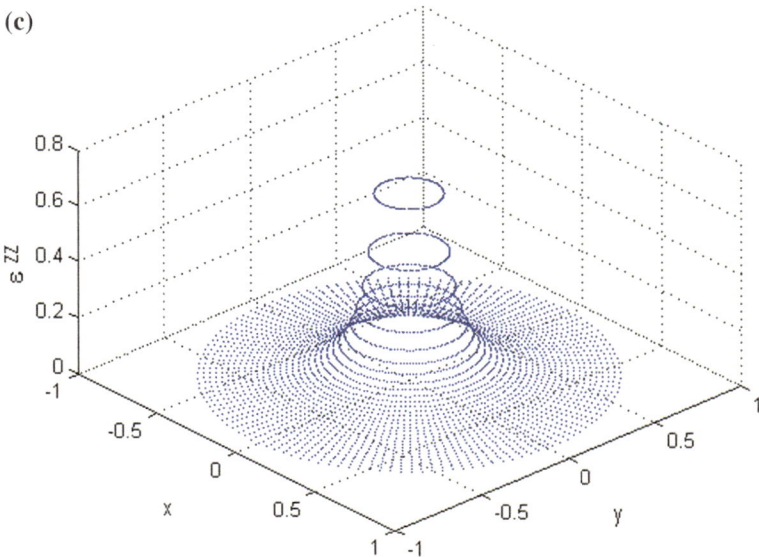

Fig. 19 (continued)

5 Permittivity Tensor for Cloaking Shell of Ogive Structure

Ogive is an extension of orthogonal coordinate system. The radome for various combat aircraft structures are of ogive shape. The parametric equations of a ogive is given as

$$x = \frac{a * \sin u * \cos \varphi}{\cosh v - \cos u}; \quad y = \frac{a * \sin u * \sin \varphi}{\cosh v - \cos u}; \quad z = \frac{a * \sinh v}{\cosh v - \cos u} \qquad (105)$$

where, u is angle varies from $0°$ to $180°$, v is from ($-\infty \leq v \leq \infty$) and φ is the angle that varies from $0°$ to $360°$. Figure 20 shows the boundary of an ogive shape generated using Matlab.

The position vector of the above mentioned parametric equation was derived (Appendix J) and the spatial metric is calculated as given below

$$\gamma_{ij} = \begin{pmatrix} \frac{1}{(\cosh v - \cos u)^2} & 0 & 0 \\ 0 & \frac{\sin^2 u}{(\cosh v - \cos u)^2} & 0 \\ 0 & 0 & \frac{1}{(\cosh v - \cos u)^2} \end{pmatrix}; \qquad (106)$$

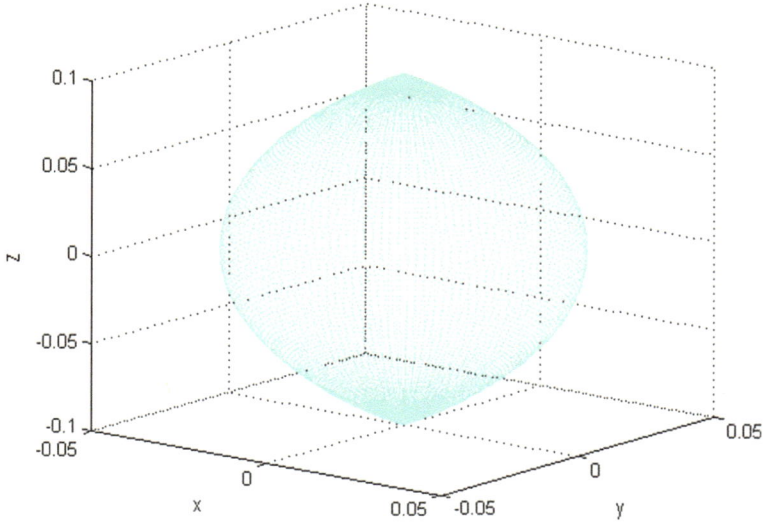

Fig. 20 Boundary of an ogive shaped cloaking shell

$$\gamma = \det(\gamma_{ij}) = \frac{\sin^2 u}{(\cosh v - \cos u)^6} \tag{107}$$

Assuming, the region where the object to be hidden $0 < u < U1$ (Fig. 20), and the region of invisibility cloaking shell is $U1 < u < U2$, then the physical medium w.r.t. the primed empty curved space-time is given by

$$u = U1 + u' * \frac{(U2 - U1)}{U2}; \quad v = v'; \quad \varphi = \varphi'; \tag{108}$$

$$\frac{\partial u}{\partial u'} = \frac{U2 - U1}{U2} \tag{109}$$

The effective geometry corresponding to the bi-anisotropic medium can be defined as

$$
\begin{aligned}
g_{ij} &= \frac{\partial x^i}{\partial x'^k} \frac{\partial x^j}{\partial x'^l} \gamma'^{kl} \\
&= \begin{pmatrix}
\left(\frac{U2-U1}{U2}\right)^2 *(\cosh v' - \cos u')^2 & 0 & 0 \\
0 & \frac{(\cosh v' - \cos u')^2}{\sin^2 u'} & 0 \\
0 & 0 & (\cosh v' - \cos u')^2
\end{pmatrix};
\end{aligned}
$$
$$\tag{110}$$

$$\sqrt{-g} = \sqrt{-g'} * \frac{\partial(x'^1, x'^2, x'^3)}{\partial(x^1, x^2, x^3)} = \left(\frac{U2}{U2 - U1}\right) * \frac{\sin u'}{(\cosh v' - \cos u')^3}; \quad (111)$$

$$\sqrt{\gamma} = \frac{\sin u}{(\cosh v - \cos u)^3} \quad (112)$$

The permittivity tensors are given by

$$\varepsilon^{ij} = \pm \frac{\sqrt{-g}}{\sqrt{\gamma}} * g^{ij} \quad (113)$$

Using Eqs. (109) through (112), the final expression for permittivity tensors can be derived as:

$$\varepsilon^{ij} = \frac{\left(\frac{U2}{U2 - U1}\right) * \frac{\sin u'}{(\cosh v' - \cos u')^3}}{\frac{\sin u}{(\cosh v - \cos u)^3}} *$$

$$\begin{pmatrix} \left(\frac{U2 - U1}{U2}\right)^2 * (\cosh v' - \cos u')^2 & 0 & 0 \\ 0 & \frac{(\cosh v' - \cos u')^2}{\sin^2 u'} & 0 \\ 0 & 0 & (\cosh v' - \cos u')^2 \end{pmatrix}$$

$$= \frac{(\cosh v - \cos u)^3}{\sin u} *$$

$$\begin{pmatrix} \left(\frac{U2 - U1}{U2}\right) * \frac{\sin u'}{(\cosh v' - \cos u')} & 0 & 0 \\ 0 & \left(\frac{U2}{U2 - U1}\right) * \frac{1}{\sin u' * (\cosh v' - \cos u')} & 0 \\ 0 & 0 & \left(\frac{U2}{U2 - U1}\right) * \frac{\sin u'}{(\cosh v' - \cos u')} \end{pmatrix}$$

$$(114)$$

The permittivity tensor derived (Eq. 113) is used for simulation of typical ogive shaped cloaking shell having dimensions $u = 5 * pi/4$, $a = 0.1$, $U1 = 0.1$, $U2 = 0.2$, and v varies from -10 to 10. To hide an object of dimension $0 < u < U1$, the permittivity tensors in the uu, vv and zz directions are given in Fig. 21a through c.

6   Conclusion

This work opens up a new possibility for calculation of permittivity and permeability tensors for various shapes of cloaking shells. The parametric equations for all the quadric surfaces generated by the eleven Eisenhart coordinate system are considered for calculation of position vectors and then the spatial metric, which

(a)

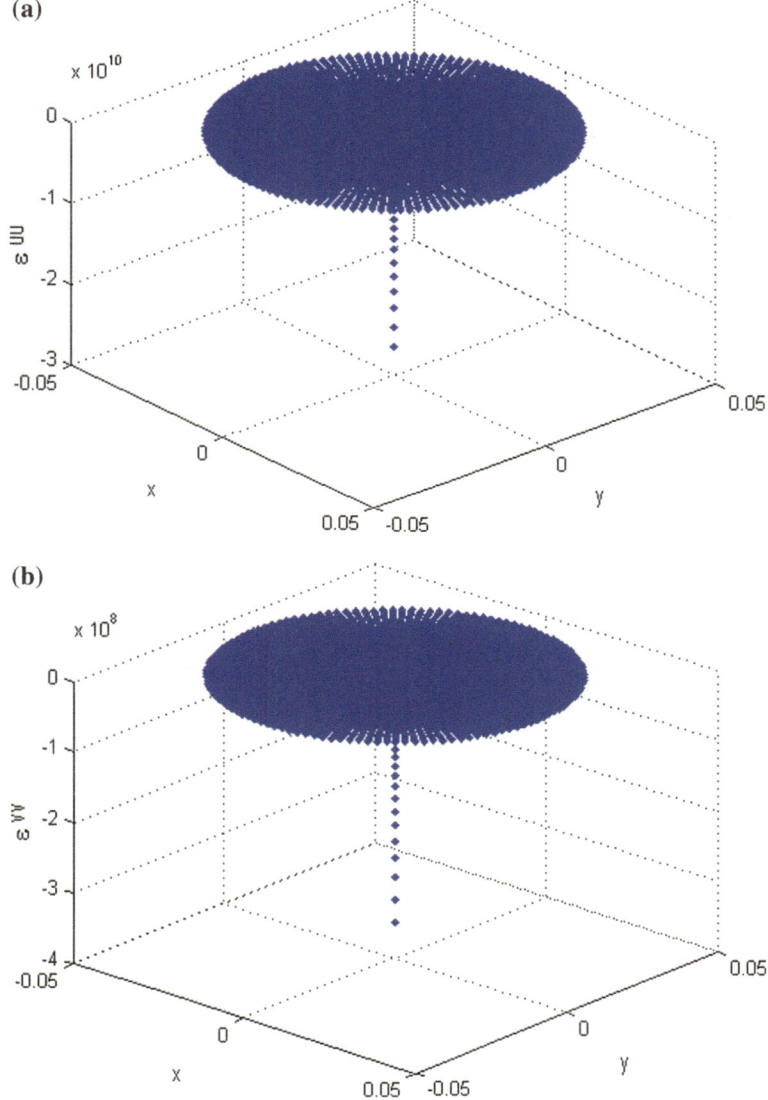

Fig. 21 a Permittivity tensor ε^{uu} distribution inside the invisibility cloaking shell of ogive shape.
b Permittivity tensor ε^{vv} distribution inside the invisibility cloaking shell of ogive shape.
c Permittivity tensor $\varepsilon^{\varphi\varphi}$ distribution inside the invisibility cloaking shell of ogive shape

(c)

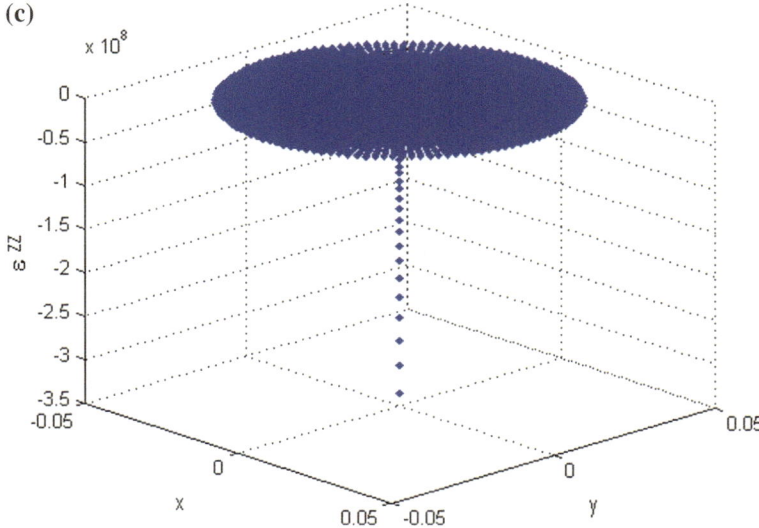

Fig. 21 (continued)

is the foremost requirement for calculation of the permittivity tensors. The mathematical derivations and simulation results for the variation of permittivity tensors in all the directions (*uu*, *vv*, and *zz*) w.r.t. the quadric surfaces are reported in the paper. The quadric cylinders, quadric surface of revolutions and ogive structures are considered for simulation as they are the important structures for modeling in aerospace platform. These derivations can be used for design of cloaking devices of various hybrid structures. It can be observed that for getting invisibility one has to go for zero refractive index and negative refractive index materials which is possible by metamaterial science and technology.

References

Ahn, D. 2006. Calculation of permittivity tensors for invisibility devices by effective medium approach in general relativity. *Document Submitted to Center for Quantum Information Processing.* Department of Electrical and Computer Engineering, University of Seoul, Seoul 130–743, Republic of Korea, pp. 1–40.

Cai, W., U.K. Chettiar, A.V. Kildishev, and V.M. Shalaev. 2007. Optical cloaking with metamaterials. *Nature Photonics* 1: 224–227.

Choudhury, B., R.M. Jha. 2013. A review of metamaterial invisibility cloak. *Computers, Materials & Continua,* 16p. (In Press).

Choudhury, B., S. Bisoyi, R.M. Jha. 2012. Emerging trends in soft computing for metamaterial design and optimization. *Computers, Materials & Continua,* 31(3): 201–228.

Jha, R.M., and W. Wiesbeck. 1995. Geodesic constant method: A novel approach to analytical surface-ray tracing on convex conducting bodies. *IEEE Antennas and Propagation Magazine* 37: 23–38.

Lee, C.H., and J. Lee. 2012. Modal characteristics of five layered slab waveguides with double clad metamaterials. *Computers, Materials & Continua* 31(2): 147–156.

Leonhardt, U. 2006. Optical conformal mapping. *Science* 312: 1777–1780.

Leonhardt, U., and T.G. Philbin. 2006. General relativity in electrical engineering. *New Journal of Physics* 8: 1–18.

Leonhardt, U., and T. Tyc. 2009. Broadband invisibility by non-euclidean cloaking. *Science* 323: 1–3.

Ma, H. 2008. Material parameter equation for elliptical cylindrical cloaks. *Science* 312: 1–4.

Plebanski, J. 1960. Electromagnetic Waves in Gravitational Fields. *Physical Review* 118(5): 1396–1408.

Ren, C., Z. Xiang, Z. Cen. 2011. The design of 2D isotropic acoustic metamaterials. International Conference on Computational and Experimental Engineering and Sciences, ICCES'11, vol. 16, no. 4, pp. 121.

Schurig, D., J.J. Mock, B.J. Justice, S.A. Cummer, J.B. Pendry, A.F. Starr, and D.R. Smith. 2006. Metamaterial electromagnetic cloak at microwave frequencies. *Science Express* 314: 977–979.

Xu, X., Y. Feng, Z. Yu, T. Jiang, and J. Zhao. 2010. Simplified ground plane invisibility cloak by multilayer dielectrics. *Optics Express* 18(24): 24477–24485.

Zhang, J., H. Jiangtao, L.Yu, H. Chen, J.A. Kong, and B.I. Wu. 2008. Cloak for multilayered and gradually changing media. *Physical Review* B 77: 035116(1)–035116(5).

Appendix A
Spatial Metric Derivations for Right Circular Cylinder

The position vector can be expressed as

$$\overrightarrow{p} = (u \cos v)e1 + (u \sin v)e2 + ze3$$

The derivates of this position vector are calculated as

$$\overrightarrow{E1} = \frac{\partial p}{\partial u} = \cos v e1 + \sin v e2$$

$$\overrightarrow{E2} = \frac{\partial p}{\partial v} = u(-\sin v)e1 + (u \cos v)e2$$

$$\overrightarrow{E3} = \frac{\partial p}{\partial z} = e3$$

The magnitudes are given by

$$|E1| = \overline{)(\cos v)^2 + (\sin v)^2}$$

$$|E1| = 1$$

$$|E2| = \overline{)(-u \sin v)^2 + (u \cos v)^2}$$

$$|E2| = \sqrt{u^2 (\cos^2 v + \sin^2 v)}$$

$$|E2| = \sqrt{u^2}$$

$$|E3| = \overline{)1^2}$$

$$|E3| = 1$$

© The Author(s) 2016
B. Choudhury et al., *Permittivity and Permeability Tensors for Cloaking Applications*, SpringerBriefs in Computational Electromagnetics, DOI 10.1007/978-981-287-805-2

Appendix B
Spatial Metric Derivations for Elliptic Cylinder

The position vector can be expressed as

$$\vec{p} = (\cosh u * \cos v)e1 + (\sinh u * \sin v)e2 + ze3$$

The derivates of this position vector are calculated as

$$\vec{E1} = \frac{\partial p}{\partial u} = (\sinh u * \cos v)e1 + (\cosh u * \sin v)e2$$

$$\vec{E2} = \frac{\partial p}{\partial v} = (-\cosh u * \sin v)e1 + (\sinh u * \cos v)e2$$

$$\vec{E3} = \frac{\partial p}{\partial z} = e3$$

The magnitudes are given by

$$|E1| = \sqrt{(-\sinh u * \cos v)^2 + (\cosh u * \sin v)^2}$$

$$|E1| = \sqrt{\sinh^2 u * \cos^2 v + \cosh^2 u * \sin^2 v}$$

$$|E1| = \sqrt{\sinh^2 u * \cos^2 v + (1 + \sinh^2 u) * \sin^2 v}$$

$$|E1| = \sqrt{\sinh^2 u * (\cos^2 v + \sin^2 v) + (\sin^2 v)}$$

$$|E1| = \sqrt{\sinh^2 u + \sin^2 v}$$

$$|E2| = \sqrt{\sinh^2 u * (\cos^2 v + \sin^2 v) + (\sin^2 v)}$$

$$|E2| = \sqrt{\sinh^2 u * \cos^2 v + (1 + \sinh^2 u) * \sin^2 v}$$

© The Author(s) 2016
B. Choudhury et al., *Permittivity and Permeability Tensors for Cloaking Applications*, SpringerBriefs in Computational Electromagnetics, DOI 10.1007/978-981-287-805-2

$$|E2| = \overline{)\sinh^2 u * \cos^2 v + \left(1 + \sinh^2 u\right) * \sin^2 v}$$

$$|E2| = \overline{)\sinh^2 u * \cos^2 v + \sin^2 v + \left(\sinh^2 u * \sin^2 v\right)}$$

$$|E2| = \overline{)\sinh^2 u * \left(\cos^2 v + \sin^2 v\right) + \left(\sin^2 v\right)}$$

$$|E2| = \overline{)\sinh^2 u + \sin^2 v}$$

$$|E3| = 1$$

Appendix C
Spatial Metric Derivations for Hyperbolic Cylinder

The position vector can be expressed as

$$\overrightarrow{p} = (\cosh u * \cos v)e1 + (\sinh u * \sin v)e2 + ze3$$

The derivates of this position vector are calculated as

$$\overrightarrow{E1} = \frac{\partial p}{\partial u} = (\sinh u * \cos v)e1 + (\cosh u * \sin v)e2$$

$$\overrightarrow{E2} = \frac{\partial p}{\partial v} = (-\cosh u * \sin v)e1 + (\sinh u * \cos v)e2$$

$$\overrightarrow{E3} = \frac{\partial p}{\partial z} = e3$$

The magnitudes are given by

$$|E1| = \sqrt{(-\sinh u * \cos v)^2 + (\cosh u * \sin v)^2}$$

$$|E1| = \sqrt{\sinh^2 u * \cos^2 v + \cosh^2 u * \sin^2 v}$$

$$|E1| = \sqrt{\sinh^2 u * \cos^2 v + (1 + \sinh^2 u) * \sin^2 v}$$

$$|E1| = \sqrt{\sinh^2 u * \cos^2 v + \sin^2 v + (\sinh^2 u * \sin^2 v)}$$

$$|E1| = \sqrt{\sinh^2 u * (\cos^2 v + \sin^2 v) + (\sin^2 v)}$$

$$|E1| = \sqrt{\sinh^2 u + \sin^2 v}$$

$$|E2| = \sqrt{\sinh^2 u * (\cos^2 v + \sin^2 v) + (\sin^2 v)}$$

© The Author(s) 2016
B. Choudhury et al., *Permittivity and Permeability Tensors for Cloaking Applications*, SpringerBriefs in Computational Electromagnetics, DOI 10.1007/978-981-287-805-2

$$|E2| = \overline{)\sinh^2 u * \cos^2 v + \cosh^2 u * \sin^2 v}$$

$$|E2| = \overline{)\sinh^2 u * \cos^2 v + \left(1 + \sinh^2 u\right) * \sin^2 v}$$

$$|E2| = \overline{)\sinh^2 u * \cos^2 v + \left(1 + \sinh^2 u\right) * \sin^2 v}$$

$$|E2| = \overline{)\sinh^2 u * \cos^2 v + \sin^2 v + \left(\sinh^2 u * \sin^2 v\right)}$$

$$|E2| = \overline{)\sinh^2 u * \left(\cos^2 v + \sin^2 v\right) + \left(\sin^2 v\right)}$$

$$|E2| = \overline{)\sinh^2 u + \sin^2 v}$$

$$|E3| = 1$$

Appendix D
Spatial Metric Derivations for Parabolic Cylinder

The position vector can be expressed as

$$\vec{p} = \frac{1}{2}\left(u^2 - v^2\right)e1 + uve2 + ze3$$

The derivates of this position vector are calculated as

$$\overrightarrow{E1} = \frac{\partial p}{\partial u} = ue1 + ve2$$

$$\overrightarrow{E2} = \frac{\partial p}{\partial v} = -ve1 + ue3$$

$$\overrightarrow{E3} = \frac{\partial p}{\partial z} = e3$$

The magnitudes are given by

$$|E1| = \sqrt{u^2 + v^2}$$

$$|E2| = \sqrt{u^2 + (-v)^2}$$

$$|E2| = \overline{)v^2 + u^2}$$

$$|E3| = 1$$

© The Author(s) 2016
B. Choudhury et al., *Permittivity and Permeability
Tensors for Cloaking Applications*, SpringerBriefs in Computational
Electromagnetics, DOI 10.1007/978-981-287-805-2

Appendix E
Spatial Metric Derivations for Sphere

The position vector can be expressed as

$$\vec{p} = (u * \sin v * \cos \varphi)e1 + (u * \sin v * \sin \varphi)e2 + (u * \cos v)e3$$

The derivates of this position vector are calculated as

$$\overrightarrow{E1} = \frac{\partial p}{\partial u} = (\sin v * \cos \varphi)e1 + (\sin v * \sin \varphi)e2 + (\cos v)e3$$

$$\overrightarrow{E2} = \frac{\partial p}{\partial v} = (u * \cos v * \cos \varphi)e1 + (u * \cos v * \sin \varphi)e2 + (-u * \sin v)e3$$

$$\overrightarrow{E3} = \frac{\partial p}{\partial \varphi} = (-u * \sin v * \sin \varphi)e1 + (u * \sin v * \cos \varphi)e2$$

The magnitudes are given by

$$|E1| = \sqrt{(\sin v * \cos \varphi)^2 + (\sin v * \sin \varphi)^2 + (\cos v)^2}$$

$$|E1| = \sqrt{(\sin^2 v * \cos^2 \varphi) + (\sin^2 v * \sin^2 \varphi) + (\cos^2 v)}$$

$$|E1| = \sqrt{(\sin^2 v) * (\cos^2 \varphi + \sin^2 \varphi) + (\cos^2 v)}$$

$$|E1| = \sqrt{(\sin^2 v) * (\cos^2 \varphi + \sin^2 \varphi) + (\cos^2 v)}$$

$$|E1| = \sqrt{(\sin^2 v) + (\cos^2 v)}$$

$$|E1| = 1$$

$$|E2| = \sqrt{(u * \cos v * \cos \varphi)^2 + (u * \cos v * \sin \varphi)^2 + (-u * \sin v)^2}$$

© The Author(s) 2016
B. Choudhury et al., *Permittivity and Permeability
Tensors for Cloaking Applications*, SpringerBriefs in Computational
Electromagnetics, DOI 10.1007/978-981-287-805-2

$$|E2| = \sqrt{\left(u^2 * \cos^2 v * \cos^2 \varphi\right) + \left(u^2 * \cos^2 v * \sin^2 \varphi\right) + \left(u^2 * \sin^2 v\right)}$$

$$|E2| = \sqrt{\left(u^2 * \cos^2 v\right) * \left(\cos^2 \varphi + \sin^2 \varphi\right) + \left(u^2 * \sin^2 v\right)}$$

$$|E2| = \sqrt{\left(u^2 * \cos^2 v\right) + \left(u^2 * \sin^2 v\right)}$$

$$|E2| = \sqrt{u^2 * \left(\cos^2 v + \sin^2 v\right)}$$

$$|E2| = \sqrt{u^2}$$

$$|E2| = u$$

$$|E3| = \sqrt{(-u * \sin v * \sin \varphi)^2 + (u * \sin v * \cos \varphi)^2}$$

$$|E3| = \sqrt{\left(u^2 * \sin^2 v * \sin^2 \varphi\right) + \left(u^2 * \sin^2 v * \cos^2 \varphi\right)}$$

$$|E3| = \sqrt{\left(u^2 * \sin^2 v\right) * \left(\sin^2 \varphi + \cos^2 \varphi\right)}$$

$$|E3| = \sqrt{\left(u^2 * \sin^2 v\right)}$$

$$|E3| = u * \sin v$$

Appendix F
Spatial Metric Derivations for Cone

The position vector can be expressed as

$$\overrightarrow{p} = (u * \sin v * \cos \varphi)e1 + (u * \sin v * \sin \varphi)e2 + (u * \cos v)e3$$

The derivates of this position vector are calculated as

$$\overrightarrow{E1} = \frac{\partial p}{\partial u} = (\sin v * \cos \varphi)e1 + (\sin v * \sin \varphi)e2 + (\cos v)e3$$

$$\overrightarrow{E2} = \frac{\partial p}{\partial v} = (u * \cos v * \cos \varphi)e1 + (u * \cos v * \sin \varphi)e2 + (-u * \sin v)e3$$

$$\overrightarrow{E3} = \frac{\partial p}{\partial \varphi} = (-u * \sin v * \sin \varphi)e1 + (u * \sin v * \cos \varphi)e2$$

The magnitudes are given by

$$|E1| = \sqrt{(\sin v * \cos \varphi)^2 + (\sin v * \sin \varphi)^2 + (\cos v)^2}$$

$$|E1| = \sqrt{(\sin^2 v * \cos^2 \varphi) + (\sin^2 v * \sin^2 \varphi) + (\cos^2 v)}$$

$$|E1| = \sqrt{(\sin^2 v) * (\cos^2 \varphi + \sin^2 \varphi) + (\cos^2 v)}$$

$$|E1| = \sqrt{(\sin^2 v) * (\cos^2 \varphi + \sin^2 \varphi) + (\cos^2 v)}$$

$$|E1| = \sqrt{(\sin^2 v) + (\cos^2 v)}$$

$$|E1| = 1$$

$$|E2| = \sqrt{(u * \cos v * \cos \varphi)^2 + (u * \cos v * \sin \varphi)^2 + (-u * \sin v)^2}$$

© The Author(s) 2016
B. Choudhury et al., *Permittivity and Permeability*
Tensors for Cloaking Applications, SpringerBriefs in Computational
Electromagnetics, DOI 10.1007/978-981-287-805-2

$$|E2| = \sqrt{\left(u^2 * \cos^2 v * \cos^2 \varphi\right) + \left(u^2 * \cos^2 v * \sin^2 \varphi\right) + \left(u^2 * \sin^2 v\right)}$$

$$|E2| = \sqrt{\left(u^2 * \cos^2 v\right) * \left(\cos^2 \varphi + \sin^2 \varphi\right) + \left(u^2 * \sin^2 v\right)}$$

$$|E2| = \sqrt{\left(u^2 * \cos^2 v\right) + \left(u^2 * \sin^2 v\right)}$$

$$|E2| = \sqrt{u^2 * \left(\cos^2 v + \sin^2 v\right)}$$

$$|E2| = \sqrt{u^2}$$

$$|E2| = u$$

$$|E3| = \sqrt{\left(-u * \sin v * \sin \varphi\right)^2 + \left(u * \sin v * \cos \varphi\right)^2}$$

$$|E3| = \sqrt{\left(u^2 * \sin^2 v * \sin^2 \varphi\right) + \left(u^2 * \sin^2 v * \cos^2 \varphi\right)}$$

$$|E3| = \sqrt{\left(u^2 * \sin^2 v\right) * \left(\sin^2 \varphi + \cos^2 \varphi\right)}$$

$$|E3| = \sqrt{\left(u^2 * \sin^2 v\right)}$$

$$|E3| = u * \sin v$$

Appendix G
Spatial Metric Derivations for Prolate Spheroid

The position vector can be expressed as

$$\vec{p} = (\sinh u * \sin v * \cos \varphi)e1 + (\sinh u * \sin v * \sin \varphi)e2 + (\cosh u * \cos v)e3$$

The derivates of this position vector are calculated as

$$\vec{E1} = \frac{\partial p}{\partial u}$$
$$= (\cosh u * \sin v * \cos \varphi)e1 + (\cosh u * \sin v * \sin \varphi)e2 + (\sinh u * \cos v)e3$$

$$\vec{E2} = \frac{\partial p}{\partial v}$$
$$= (\sinh u * \cos v * \cos \varphi)e1 + (\sinh u * \cos v * \sin \varphi)e2 + (-\cosh u * \sin v)e3$$

$$\vec{E3} = \frac{\partial p}{\partial \varphi} = (-\sinh u * \sin v * \sin \varphi)e1 + (\sinh u * \sin v * \cos \varphi)e2$$

The magnitude of $E1$ is given by

$$|E1| = \sqrt{(\cosh u * \sin v * \cos \varphi)^2 + (\cosh u * \sin v * \sin \varphi)^2 + (\sinh u * \cos v)^2}$$

$$|E1| = \sqrt{(\cosh^2 u * \sin^2 v * \cos^2 \varphi) + (\cosh^2 u * \sin^2 v * \sin^2 \varphi) + (\sinh^2 u * \cos^2 v)}$$

$$|E1| = \sqrt{(\cosh^2 u * \sin^2 v) * (\cos^2 \varphi + \sin^2 \varphi) + (\sinh^2 u * \cos^2 v)}$$

$$|E1| = \sqrt{(\cosh^2 u * \sin^2 v) * (\cos^2 \varphi + \sin^2 \varphi) + (\sinh^2 u * \cos^2 v)}$$

$$|E1| = \sqrt{(\cosh^2 u * \sin^2 v) + (\sinh^2 u * \cos^2 v)}$$

$$|E1| = \sqrt{\sinh^2 u * \cos^2 v + (1 + \sinh^2 u) * \sin^2 v}$$

© The Author(s) 2016
B. Choudhury et al., *Permittivity and Permeability
Tensors for Cloaking Applications*, SpringerBriefs in Computational
Electromagnetics, DOI 10.1007/978-981-287-805-2

$$|E1| = \sqrt{\sinh^2 u * \cos^2 v + \sin^2 v + \left(\sinh^2 u * \sin^2 v\right)}$$

$$|E1| = \sqrt{\sinh^2 u * \left(\cos^2 v + \sin^2 v\right) + \left(\sin^2 v\right)}$$

$$|E1| = \sqrt{\sinh^2 u + \sin^2 v}$$

The magnitude of *E2* is given by

$$|E2| = \sqrt{(\sinh u * \cos v * \cos \varphi)^2 + (\sinh u * \cos v * \sin \varphi)^2 + (-\cosh u * \sin v)^2}$$

$$|E2| = \sqrt{\left(\sinh^2 u * \cos^2 v * \cos^2 \varphi\right) + \left(\sinh^2 u * \cos^2 v * \sin^2 \varphi\right) + \left(\cosh^2 u * \sin^2 v\right)}$$

$$|E2| = \sqrt{\left(\sinh^2 u * \cos^2 v\right) * \left(\cos^2 \varphi + \sin^2 \varphi\right) + \left(\cosh^2 u * \sin^2 v\right)}$$

$$|E2| = \sqrt{\left(\sinh^2 u * \cos^2 v\right) + \left(\cosh^2 u * \sin^2 v\right)}$$

$$|E2| = \sqrt{\sinh^2 u * \cos^2 v + \left(1 + \sinh^2 u\right) * \sin^2 v}$$

$$|E2| = \sqrt{\sinh^2 u * \cos^2 v + \sin^2 v + \left(\sinh^2 u * \sin^2 v\right)}$$

$$|E2| = \sqrt{\sinh^2 u * \left(\cos^2 v + \sin^2 v\right) + \left(\sin^2 v\right)}$$

$$|E2| = \sqrt{\sinh^2 u + \sin^2 v}$$

The magnitude of *E3* is given by

$$|E3| = \sqrt{(-\sinh u * \sin v * \sin \varphi)^2 + (\sinh u * \sin v * \cos \varphi)^2}$$

$$|E3| = \sqrt{\left(\sinh^2 u * \sin^2 v * \sin^2 \varphi\right) + \left(\sinh^2 u * \sin^2 v * \cos^2 \varphi\right)}$$

$$|E3| = \sqrt{\left(\sinh^2 u * \sin^2 v\right) * \left(\sin^2 \varphi + \cos^2 \varphi\right)}$$

$$|E3| = \sqrt{\left(\sinh^2 u * \sin^2 v\right)}$$

Appendix H
Spatial Metric Derivations for Oblate Spheroid

The position vector can be expressed as

$$\vec{p} = (\cosh u * \cos v * \cos \varphi)e1 + (\cosh u * \cos v * \sin \varphi)e2 + (\sinh u * \sin v)e3$$

The derivates of this position vector are calculated as

$$\vec{E1} = \frac{\partial p}{\partial u}$$
$$= (\sinh u * \cos v * \cos \varphi)e1 + (\sinh u * \cos v * \sin \varphi)e2 + (\cosh u * \sin v)e3$$

$$\vec{E2} = \frac{\partial p}{\partial v}$$
$$= (-\cosh u * \sin v * \cos \varphi)e1 + (-\cosh u * \sin v * \sin \varphi)e2$$
$$+ (\sinh u * \cos v)e3$$

$$\vec{E3} = \frac{\partial p}{\partial \varphi} = (\cosh u * \cos v * \sin \varphi)e1 + (\cosh u * \cos v * \cos \varphi)e2$$

The magnitude of $E1$ given by

$$|E1| = \overline{)(\sinh u * \cos v * \cos \varphi)^2 + (\sinh u * \cos v * \sin \varphi)^2 + (\cosh u * \sin v)^2}$$

$$|E1| = \overline{)\,(\sinh^2 u * \cos^2 v) * (\cos^2 \varphi + \sin^2 \varphi) + (\cosh^2 u * \sin^2 v)}$$

$$|E1| = \overline{)\,(\sinh^2 u * \cos^2 v) * (\cos^2 \varphi + \sin^2 \varphi) + (\cosh^2 u * \sin^2 v)}$$

$$|E1| = \overline{)\,(\cosh^2 u * \sin^2 v) + (\sinh^2 u * \cos^2 v)}$$

$$|E1| = \overline{)\,\sinh^2 u * \cos^2 v + (1 + \sinh^2 u) * \sin^2 v}$$

$$|E1| = \overline{)\,\sinh^2 u * \cos^2 v + \sin^2 v + (\sinh^2 u * \sin^2 v)}$$

© The Author(s) 2016
B. Choudhury et al., *Permittivity and Permeability*
Tensors for Cloaking Applications, SpringerBriefs in Computational
Electromagnetics, DOI 10.1007/978-981-287-805-2

$$|E1| = \sqrt{\sinh^2 u * \left(\cos^2 v + \sin^2 v\right) + \left(\sin^2 v\right)}$$

$$|E1| = \sqrt{\sinh^2 u + \sin^2 v}$$

The magnitude of $E2$ is given by

$$|E2| = \sqrt{\left(-\cosh u * \sin v * \cos \varphi\right)^2 + \left(-\cosh u * \sin v * \sin \varphi\right)^2 + \left(\sinh u * \cos v\right)^2}$$

$$|E2| = \sqrt{\left(\cosh^2 u * \sin^2 v * \cos^2 \varphi\right) + \left(\cosh^2 u * \sin^2 v * \sin^2 \varphi\right) + \left(\sinh^2 u * \cos^2 v\right)}$$

$$|E2| = \sqrt{\left(\cosh^2 u * \sin^2 v\right) * \left(\cos^2 \varphi + \sin^2 \varphi\right) + \left(\sinh^2 u * \cos^2 v\right)}$$

$$|E2| = \sqrt{\left(\sinh^2 u * \cos^2 v\right) + \left(\cosh^2 u * \sin^2 v\right)}$$

$$|E2| = \sqrt{\sinh^2 u * \cos^2 v + \left(1 + \sinh^2 u\right) * \sin^2 v}$$

$$|E2| = \sqrt{\sinh^2 u * \cos^2 v + \sin^2 v + \left(\sinh^2 u * \sin^2 v\right)}$$

$$|E2| = \sqrt{\sinh^2 u * \left(\cos^2 v + \sin^2 v\right) + \left(\sin^2 v\right)}$$

$$|E2| = \sqrt{\sinh^2 u + \sin^2 v}$$

The magnitude of $E3$ is given by

$$|E3| = \sqrt{\left(\cosh u * \cos v * \sin \varphi\right)^2 + \left(\cosh u * \cos v * \cos \varphi\right)^2}$$

$$|E3| = \sqrt{\left(\cosh^2 u * \cos^2 v * \sin^2 \varphi\right) + \left(\cosh^2 u * \cos^2 v * \cos^2 \varphi\right)}$$

$$|E3| = \sqrt{\left(\cosh^2 u * \cos^2 v\right) * \left(\sin^2 \varphi + \cos^2 \varphi\right)}$$

$$|E3| = \sqrt{\left(\cosh^2 u * \cos^2 v\right)}$$

Appendix I
Spatial Metric Derivations for GPOR

The position vector can be expressed as

$$\overrightarrow{p} = (uv\cos\varphi)e1 + (uv\sin\varphi)e2 - \frac{1}{2} * (u^2 - v^2)e3$$

The derivates of this position vector are calculated as

$$\overrightarrow{E1} = \frac{\partial p}{\partial u} = (v * \cos\varphi)e1 + (v * \sin\varphi)e2 - (u)e3$$

$$\overrightarrow{E2} = \frac{\partial p}{\partial v} = (u * \cos\varphi)e1 + (u * \sin\varphi)e2 + ve3$$

$$\overrightarrow{E3} = \frac{\partial p}{\partial \varphi} = (-u * v * \sin\varphi)e1 + (u * v * \cos\varphi)e2$$

The magnitudes are given by

$$|E1| = \sqrt{(u * \cos\varphi)^2 + (u * \sin\varphi)^2 + (-v)^2}$$

$$|E1| = \sqrt{u^2 + (v)^2}$$

$$|E2| = \sqrt{(v * \cos\varphi)^2 + (v * \sin\varphi)^2 + (u)^2}$$

$$|E2| = \sqrt{u^2 + v^2}$$

$$|E3| = \sqrt{(v * u * \cos\varphi)^2 + (-v * u * \sin\varphi)^2}$$

$$|E3| = uv$$

© The Author(s) 2016
B. Choudhury et al., *Permittivity and Permeability Tensors for Cloaking Applications*, SpringerBriefs in Computational Electromagnetics, DOI 10.1007/978-981-287-805-2

Appendix J
Spatial Metric Derivations for Ogive

The position vector can be expressed as

$$\vec{p} = \left(\frac{\sin u * \cos \varphi}{\cosh v - \cos u}\right) e1 + \left(\frac{\sin u * \sin \varphi}{\cosh v - \cos u}\right) e2 + \left(\frac{\sinh v}{\cosh v - \cos u}\right) e3$$

The derivates of this position vector are calculated as

$$\vec{E1} = \frac{\partial p}{\partial u}$$
$$= \frac{\cos \varphi * (\cosh v * \cos u - \cos^2 u) - \sin^2 u}{(\cosh v - \cos u)^2} e1 + \frac{\sin \varphi * (\cosh v * \cos u - \cos^2 u) - \sin^2 u}{(\cosh v - \cos u)^2} e2 - \frac{\sin u * \sinh v}{(\cosh v - \cos u)^2} e3$$

$$\vec{E2} = \frac{\partial p}{\partial \varphi} = \left(\frac{-\sin u * \sin \varphi}{(\cosh v - \cos u)^2}\right) e1 + \left(\frac{\sin u * \cos \varphi}{(\cosh v - \cos u)^2}\right) e2$$

$$\vec{E3} = \frac{\partial p}{\partial n} = \frac{-\sin u * \cos \varphi * \sinh v}{(\cosh v - \cos u)^2} + \frac{-\sin u * \sin \varphi * \sinh v}{(\cosh v - \cos u)^2} + \frac{\cosh^2 v - (\cos u * \cosh v) - \sinh^2 n}{(\cosh v - \cos u)^2}$$

The magnitudes are given by

$$|E1| = \frac{\cos \varphi * ((\cosh v * \cos u) - (\cos^2 u + \sin^2 u))e1 + \sin \varphi * ((\cosh v * \cos u) - (\cos^2 u + \sin^2 u))e2 + (-\sin u * \sinh v)e3}{(\cosh v - \cos u)^2}$$

$$|E1| = \frac{\cos \varphi * ((\cosh v * \cos u) - 1)e1 + \sin \varphi * ((\cosh v * \cos u) - 1)e2 + (-\sin u * \sinh v)e3}{(\cosh v - \cos u)^2}$$

$$|E1| = \sqrt{\frac{\cos^2 \varphi * ((\cosh v * \cos u) - 1)^2 + \sin^2 \varphi * ((\cosh v * \cos u) - 1)^2 + (\sin^2 u * \sinh^2 v)}{(\cosh v - \cos u)^4}}$$

© The Author(s) 2016
B. Choudhury et al., *Permittivity and Permeability
Tensors for Cloaking Applications*, SpringerBriefs in Computational
Electromagnetics, DOI 10.1007/978-981-287-805-2

$$|E1| = \sqrt{\frac{(\cos^2\varphi + \sin^2\varphi) * ((\cosh v * \cos u) - 1)^2 + (\sin^2 u * \sinh^2 v)}{(\cosh v - \cos u)^4}}$$

$$|E1| = \sqrt{\frac{((\cosh v * \cos u) - 1)^2 + (\sin^2 u * \sinh^2 v)}{(\cosh v - \cos u)^4}}$$

$$|E1| = \sqrt{\frac{(\cosh^2 v * \cos^2 u + 1 - (2 * \cosh v * \cos u)) + (\sin^2 u * \sinh^2 v)}{(\cosh v - \cos u)^4}}$$

$$|E1| = \sqrt{\frac{((1 + \sinh^2 v) * \cos^2 u + 1 - (2 * \cosh v * \cos u)) + (\sin^2 u * \sinh^2 v)}{(\cosh v - \cos u)^4}}$$

$$|E1| = \sqrt{\frac{((\cos^2 u + (\sinh^2 v * \cos^2 u)) + 1 - (2 * \cosh v * \cos u)) + (\sin^2 u * \sinh^2 v)}{(\cosh v - \cos u)^4}}$$

$$|E1| = \sqrt{\frac{\cos^2 u + \sinh^2 v(\cos^2 u + \sin^2 u) - 2 * \cos u * \cosh v + 1}{(\cosh v - \cos u)^4}}$$

$$|E1| = \sqrt{\frac{\cos^2 u + \sinh^2 v - 2 * \cos u * \cosh v + \cosh^2 v - \sinh^2 v}{(\cosh v - \cos u)^4}}$$

$$|E1| = \sqrt{\frac{\cos^2 u - 2 * \cos u * \cosh v + \cosh^2 v}{(\cosh v - \cos u)^4}}$$

$$|E1| = \sqrt{\frac{(\cosh v - \cos u)^2}{(\cosh v - \cos u)^4}}$$

$$|E1| = \frac{1}{\cosh v - \cos u}$$

The magnitude of E2 is given by

$$|E2| = \sqrt{\left(\frac{-\sin u * \sin\varphi}{(\cosh\varphi - \cos u)}\right)^2 + \left(\frac{\sin u * \cos\varphi}{\cosh\varphi - \cos u}\right)^2}$$

$$|E2| = \sqrt{\frac{\sin^2 u * \sin^2\varphi}{(\cosh\varphi - \cos u)^2} + \frac{\sin^2 u * \cos^2\varphi}{(\cosh\varphi - \cos u)^2}}$$

$$|E2| = \sqrt{\frac{\sin^2 u * \left(\sin^2 \varphi + \cos^2 \varphi\right)}{\left(\cosh \varphi - \cos u\right)^2}}$$

$$|E2| = \sqrt{\frac{\sin^2 u}{\left(\cosh v - \cos u\right)^2}}$$

$$|E2| = \frac{\sin u}{\left(\cosh v - \cos u\right)}$$

The magnitude of $E3$ is given by

$$|E3| = \frac{(-\cos \varphi * \sin u * \sinh v)e1 + (-\sin \varphi * \sin u * \sinh v)e2 + \left((\cosh v * (\cosh v - \cos u)) - \sinh^2 v\right)e3}{\left(\cosh v - \cos u\right)^2}$$

$$|E3| = \frac{(-\cos \varphi * \sin u * \sinh v)e1 + (\sin \varphi * \sin u * \sinh v)e2 + \left(\cosh^2 v - (\cos u * \cosh v) - \sinh^2 v\right)e3}{\left(\cosh v - \cos u\right)^2}$$

$$|E3| = \frac{(-\cos \varphi * \sin u * \sinh v)e1 + (\sin \varphi * \sin u * \sinh v)e2 + \left(\cosh^2 v - (\cos u * \cosh v) - \sinh^2 v\right)e3}{\left(\cosh v - \cos u\right)^2}$$

$$|E3| = \frac{(-\cos \varphi * \sin u * \sinh v) + (\sin \varphi * \sin u * \sinh v) + (1 - (\cos u * \cosh v))}{\left(\cosh v - \cos u\right)^2}$$

$$|E3| = \sqrt{\frac{(-\cos \varphi * \sin u * \sinh v)^2 + (\sin \varphi * \sin u * \sinh v)^2 + (1 - (\cos u * \cosh v))^2}{\left(\cosh v - \cos u\right)^4}}$$

$$|E3| = \sqrt{\frac{\left(\cos^2 \varphi * \sin^2 u * \sinh^2 v\right)}{\left(\cosh v - \cos u\right)^4} + \frac{\left(\sin^2 \varphi * \sin^2 u * \sinh^2 v\right)}{\left(\cosh v - \cos u\right)^4} + \frac{\left(1 - \cos u * \cosh v\right)^2}{\left(\cosh v - \cos u\right)^4}}$$

$$|E3| = \sqrt{\frac{\left(\sin^2 u * \sinh^2 v\left(\cos^2 \varphi + \sin^2 \varphi\right)\right) + (1 - (\cos u * \cosh v))^2}{\left(\cosh v - \cos u\right)^4}}$$

$$|E3| = \sqrt{\frac{((\cosh v * \cos u) - 1)^2 + \left(\sin^2 u * \sinh^2 v\right)}{\left(\cosh v - \cos u\right)^4}}$$

$$|E3| = \sqrt{\frac{\left(\cosh^2 v * \cos^2 u + 1 - (2 * \cosh v * \cos u)\right) + \left(\sin^2 u * \sinh^2 v\right)}{\left(\cosh v - \cos u\right)^4}}$$

$$|E3| = \sqrt{\frac{\left(\left(1 + \sinh^2 v\right) * \cos^2 u + 1 - (2 * \cosh v * \cos u)\right) + \left(\sin^2 u * \sinh^2 v\right)}{(\cosh v - \cos u)^4}}$$

$$|E3| = \sqrt{\frac{\left(\left(\cos^2 u + \left(\sinh^2 v * \cos^2 u\right)\right) + 1 - (2 * \cosh v * \cos u)\right) + \left(\sin^2 u * \sinh^2 v\right)}{(\cosh v - \cos u)^4}}$$

$$|E3| = \sqrt{\frac{\cos^2 u + \sinh^2 v\left(\cos^2 u + \sin^2 u\right) - 2 * \cos u * \cosh v + 1}{(\cosh v - \cos u)^4}}$$

$$|E3| = \sqrt{\frac{\cos^2 u + \sinh^2 v - 2 * \cos u * \cosh v + \cosh^2 v - \sinh^2 v}{(\cosh v - \cos u)^4}}$$

$$|E3| = \sqrt{\frac{\cos^2 u - 2 * \cos u * \cosh v + \cosh^2 v}{(\cosh v - \cos u)^4}}$$

$$|E3| = \sqrt{\frac{(\cosh v - \cos u)^2}{(\cosh v - \cos u)^4}}$$

$$|E3| = \frac{1}{\cosh v - \cos u}$$

About the Book

This book is focused on derivations of analytical expressions for stealth and cloaking applications. An optimal version of electromagnetic (EM) stealth is the design of invisibility cloak of arbitrary shapes in which the EM waves can be controlled within the cloaking shell by introducing a prescribed spatial variation in the constitutive parameters. The promising challenge in design of invisibility cloaks lies in the determination of permittivity and permeability tensors for all the layers. This book provides the detailed derivation of analytical expressions of the permittivity and permeability tensors for various quadric surfaces within the eleven Eisenhart coordinate systems. These include the cylinders and the surfaces of revolutions. The analytical modeling and spatial metric for each of these surfaces are provided along with their tensors. This mathematical formulation will help the EM designers to analyze and design of various quadratics and their hybrids, which can eventually lead to design of cloaking shells of arbitrary shapes.

© The Author(s) 2016
B. Choudhury et al., *Permittivity and Permeability Tensors for Cloaking Applications*, SpringerBriefs in Computational Electromagnetics, DOI 10.1007/978-981-287-805-2

Author Index

© The Author(s) 2016
B. Choudhury et al., *Permittivity and Permeability Tensors for Cloaking Applications*, SpringerBriefs in Computational Electromagnetics, DOI 10.1007/978-981-287-805-2

Subject Index

© The Author(s) 2016
B. Choudhury et al., *Permittivity and Permeability
Tensors for Cloaking Applications*, SpringerBriefs in Computational
Electromagnetics, DOI 10.1007/978-981-287-805-2